The Intermediate Licence
Building on the Foundation

Edited by Steve Hartley, G0FUW

With contributions by:
Alan Betts, G0HIQ
Dr John Craig, G3SGR
Edwin Taylor, G3SQX

Radio Society of Great Britain

The Intermediate Licence
Building on the Foundation

Third Edition

Edited by Steve Hartley, G0FUW

Published by the Radio Society of Great Britain, Lambda House, Cranborne Road, Potters Bar, Herts EN6 3JE.
Tel: 0870 904 7373. Fax: 0870 904 7374. Web: www.rsgb.org

First published 2003.
Second edition 2003
Third edition 2005
Reprinted 2006

Radio Society of Great Britain, 2003-5. All rights reserved. No part of this publication may be reproduced, stored in a retreival system, or transmitted, in any form or by any means, electronic, mechanical, photocopying, recording or otherwise, without the prior written permission of the Radio Society of Great Britain.

The extracts from the licence and syllabus are Crown Copyright & OfCom, and are reproduced with their permission.

ISBN 1-872309-86-0

Cover design:
Bob Ryan, 2E1EKS

Design and layout:
Steve White, G3ZVW

Cartoons:
Gary Milton, G0CUQ

Revisions of Third Edition & Production:
Mark Allgar, M1MPA

Printed in Great Britain by Latimer Ternd.

The opinions expressed in this book are those of the author and not necessarily those of the RSGB. Whilst the information presented is believed to be correct, the publishers and their agents cannot accept responsibility for consequences arising from any inaccuracies or omissions.

This book has a suporting website
http://www.rsgb.org/books/extra/intermediate.htm

Any corrections and points of clarification that have not been incorporated in this printing of the book can be found here along with any supporting material that may have become available

Contents

Worksheet	Topic	Page
1	Introduction	1
2	Soldering Skills & Safety	3
3	Conductors & Insulators	5
4	Components & Symbols	6
5	Building a Simple DC Circuit	8
6	Project Briefing & Tool Safety	9
7	Fitting a 13 Amp Plug & Electrical Safety	11
8	Multi-meters & Units of Measurement	13
9	Measuring Potential Difference	15
10	Measuring Current	16
11	Licence Conditions 1	17
12	Calculating Input Power	21
13	Operating Practices & Procedures	22
14	Measuring Resistance	25
15	Capacitors, Inductors & Tuned Circuits	27
16	Demonstrating Ohm's Law	30
17	RF Oscillators	31
18	Calibrating a VFO	33
19	Semiconductors	35
20	Using Diodes	37
21	Revision Questions 1	38
22	Transmitters	39
23	Using a Transistor as a Switch	42
24	Licence Conditions 2	43
25	Power Supplies	46
26	Other Types of Transmission	48
27	Receivers	50
28	Antenna Matching	53
29	Antenna Feeders	56
30	Antenna Gain	58
31	Propagation	60
32	RF Coax Connectors	63
33	Electromagnetic Compatibility (EMC)	65
34	Checking for Harmonics & Spurious Emissions	68
35	Good Radio Housekeeping	70
36	Revision Questions 2	72
37	The Practical Assessment	73
38	The Written Examination	74

Appendix

1	Component Symbols	75
2	Syllabus Cross Reference	76

Preface

The Novice Amateur Radio Licence was launched some ten years ago as a means of introducing newcomers into the hobby and soon afterwards I enrolled as a tutor. The next thing I knew I had run a couple of courses and I have been delivering amateur radio training ever since. I was therefore very pleased to be asked to write this book.

This book draws heavily on the decade of success enjoyed by the Novice Licence scheme and in particular the *Novice Student's Notebook* by John Case, GW4HWR. However, this is more than a new edition, it is a new book to link with the revised Intermediate syllabus. Some material has been 'brought forward' from the *RAE Manual*, and other RSGB sources have been tapped to fill in a few gaps in the earlier training material. My co-authors and I have also added some new material.

I must acknowledge the hard work and dedication shown by all concerned in helping me to produce this book; in particular Ed Taylor, G3SQX, John Craig, G3SGR, and Alan Betts, G0HIQ. I would also like to thank those who helped by proof-reading our work and offering many excellent suggestions for improvement. The book would also not have been possible without the help of RSGB headquarters staff.

I hope you enjoy your preparation for the Intermediate Licence examination and assessment. Learning should be fun!

73

Steve Hartley, G0FUW
October 2002

Introduction

Preparing for the Intermediate Licence

WELCOME TO the next stage in your self-training in amateur radio. The fact that you are reading this suggests that you have already passed the Foundation Licence examination, indeed you may have been operating with your M3 callsign for some time. No matter how long it is since you first became interested in amateur radio, this book is intended to prepare you for the Intermediate Licence practical assessment and written exam.

You may not have operated with a Foundation Licence, but you *must* have passed the Foundation examination *before* you can take the Intermediate examination. There is no requirement for you to attend a formal training course to prepare for the examination or the assessment, but that is without doubt the best way to learn if you can do so. If you cannot get to a local course, don't worry, this book has been written to enable you to learn on your own before attending the assessment and examination.

If you are studying on your own, you will need to find out where your nearest assessor is. You should contact the Radio Society of Great Britain (RSGB) as soon as you start to study, as it may take some time to organise an assessment, particularly if you are in a remote part of the country.

The Intermediate syllabus

THE INTERMEDIATE syllabus covers much the same topics as the Foundation syllabus, so you should be reasonably familiar with the basics. At this level you will be building on your Foundation knowledge to gain a better understanding of radio theory, improving your knowledge of operating techniques and learning about a few new concepts.

The components and construction topics are certainly new. They are intended to bring some of the theory to life; for example, to see that Ohm's law really does work! In almost every other area the syllabus merely requires you to know a little more or to be able to demonstrate a new skill.

Using this book

THIS BOOK DOES NOT follow the order of Ofcoms's Intermediate licence syllabus, but sets out a logical scheme of work for you and/or your tutor to follow. In the main, the practical exercises follow the relevant theory sections. You should therefore work through the worksheets in the order they appear.

There are some 38 worksheets, each representing about 30 minutes of study. Some might take you less time to read through whilst others, particularly those with practical exercises, might take longer. Two sets of revision questions are included to give you some idea of your progress, and guidance on the written and practical assessments is given towards the end of the book.

Roughly 20 hours study has been allocated for the training material, but the time for *you* to complete this course of study depends on *your* level of knowledge and experience. However, the only prior knowledge you need is that which you gained from preparing for the Foundation Licence assessments. If you have a copy of the RSGB book *Foundation Licence Now!* you might like to read it through again to refresh your memory.

Other resources

THIS BOOK IS believed to contain all the information you will need to prepare for the Intermediate Licence assessments, but there are lots of other sources of information available. For example, the Intermediate Licence terms and conditions are set out in the Ofcom booklet BR68/I, which you will receive with your licence, but the examinable parts are all covered in this book.

Ofcom publishes a whole series of leaflets on amateur radio topics and the Radio Society of Great Britain's bookshop has a host of titles available. Where these additional resources are considered to be particularly helpful, details are included in the relevant worksheets.

The complete syllabus is a useful document to have to hand. This sets out quite clearly what knowledge or skills are required for the assessments. The syllabus assessment objective numbers are quoted in the worksheets and a cross-reference index is provided at the back of this book to help you track down the learning material that supports each syllabus objective.

RSGB - FOR THESE AND MANY MORE BOOKS

www.rsgb.org/shop - Tel: 0870 904 7373

Practical exercises and assessments

UNLIKE THE Foundation Licence, the Intermediate Licence allows you to build your own transmitters from scratch. Consequently you need to be able to demonstrate some basic construction skills. For this reason the assessment includes a number of practical tasks.

If you complete all the exercises in this book you will be well placed to demonstrate your competence to the assessor. Indeed, if you are taking part in an organised course, the tutor may well sign-off the assessments as you complete the exercises. You may like to look at the worksheet on the practical assessment now, to see what is ahead of you.

Worksheet 6 explains more about the project, but please bear in mind that the time you need to complete the project will be *in addition* to the 20 hours for main worksheets.

If you have a disability that prevents you from completing the practical exercises you must inform your tutor and/or assessor. In such cases the assessment will normally be based on demonstrating knowledge of the practical activities.

After the assessments

WHEN YOU HAVE completed your training and the assessments you will need to apply for your Intermediate Licence in much the same way that you did for your Foundation Licence, assuming you have taken up a Foundation Licence. Your Intermediate assessor should be able to provide you with an application form, or at least tell you where to obtain one.

A pass in the Intermediate assessments will entitle you to apply for an Amateur Radio (Intermediate) Licence with a 2E0xxx callsign in England, 2M0xxx in Scotland, 2W0xxx in Wales etc. The additional privileges over the Foundation Licence include output power up to 50 watts, access to additional bands and the use of amateur satellites.

You can make your application as soon as you have the necessary result sheet showing a 'pass'. Alternatively, you may wish to wait until your Foundation Licence is due for renewal and upgrade it then; the choice is yours. However, you should be aware that if you already have a Foundation Licence and you take up an Intermediate Licence as well, you will have to pay an additional licence fee (unless your age exempts you from paying).

The next consideration is when to take the next step towards gaining a full licence. Again, there is no one answer. You may wish to build on your success straight away, or you may prefer to use your new callsign for a while. It is entirely up to you, but we wish you success whichever path you choose.

Soldering Skills & Safety

THE INTERMEDIATE Licence allows the construction of all kinds of amateur radio equipment, including transmitters, so it is important to know a little bit about soldering. The syllabus reflects this by requiring you to demonstrate the ability to solder and to produce a radio-related project.

This worksheet deals with the basics of soldering. Later worksheets will advise on practising the skills and there is a specific worksheet about the project.

What is soldering? (10b.1)

SOLDERING IS A method of joining metal wires and components together using a hot soldering iron to melt the solder. Soldering is a bit like welding, but at a lower temperature. *Beware!*... the soldering iron and the solder are *very* hot and can cause serious burns to the skin and eyes.

Essentially, what we do is place two pieces of metal side-by-side, heat them up and then run molten metal, the solder, over the two surfaces. The solder bonds with the two pieces of metal to form a good electrical joint, then cools to form a good mechanical joint.

Soldering irons are usually powered by electricity and are rated in Watts (W). The greater the wattage, the more heat they produce. For circuit boards a 15W iron is sufficient, but for fitting RF connectors or larger jobs a 25W model would be better. The electric element heats a small tip, and it is the tip that we use to make the joint.

What is solder made from?

SOLDER FOR radio projects normally comes in the form of a wire - the large blocks of solder used by plumbers are not really of much use to us.

The solder wire is quite heavy because it is made of lead and tin, lead being a very heavy metal. Both lead and tin are metals that melt at relatively low temperatures (although still hot enough to burn!), so solder is quite easy to work with.

Lead-based solder is being replaced by silver-based solder, which is still quite heavy, but not quite so easy to work with.

What is flux? (10b.2)

NORMALLY, INSIDE the solder wire is a substance called 'flux'. When it melts with the solder wire it helps the molten metal to flow over the surfaces of the pieces of metal that we are trying to join together. Flux also helps to remove any oxide (corrosion) from those surfaces. Again, the tubs of flux sold in plumbers' merchants are of no use to the radio amateur and it is best to buy solder wire with the flux 'built in'. This is usually marked-up as 'Rosin Cored Solder', rosin being a type of flux.

Soldering different metals (10b.3)

IT IS IMPORTANT to note that not all metals solder easily. Some metals, such as tin, copper and brass solder quite well, but aluminium and stainless steel require special techniques. Normally in amateur radio projects we are joining thin component wires to copper printed circuit boards or plated steel plugs to copper wires, so there are no major problems.

Tinning the iron (10b.4)

'BEFORE A soldering iron can be used the tip should be cleaned on a wet sponge and 'tinned'. This involves melting a small amount of solder onto the hot tip to 'wet' it. The wet solder helps to transfer the heat to the items that are to be soldered.

Quick joints, but not too quick (10b.5)

SOLDER JOINTS need to be made reasonably quickly, to avoid damage to the electrical components. Some components are made from plastic and will melt if the iron is left on them for too long.

You have been warned!

On the other hand, if you don't keep the iron on for long enough, you may end up with what is known as a 'dry joint'. This is where the solder cools before it has made a good electrical bond with the pieces being joined.

Dry joints are the most common faults in radio construction projects. They are also one of the hardest to track down, so they are best avoided in the first place. If you hold the iron in place and count slowly to three after the solder has formed the joint, that should be about right. Practice will make perfect.

Good solder joints (10d.2)

YOU WILL LEARN to recognise a 'good' solder joint quite quickly. It is generally seen as a bright hemisphere when cool. If it looks 'lumpy' and/or discoloured it

Warm the joint, introduce the solder, then take it away. Finally, remove the heat.

Well-soldered joints are bright and shiny. After the solder cools, trim off any excess component leads and wires.

may well be a dry joint, in which case you should go back and clean it up. To do this, melt the soldered joint and use a solder sucker to remove the molten solder, then make the joint again. A little extra effort at this point could save a long fault finding exercise later.

Using a stand (9a.1)

TO AVOID BEING burned, the soldering iron should be rested on a purpose-built stand when not in use. These often come with an iron when you buy one, but they can be purchased separately or made using appropriate heat resistant materials.

Most stands also incorporate a sponge, which should be kept moist and used to keep the tip of the iron clean in between joints. Again, keeping the iron clean will reduce the chance of making poor joints.

Ventilation (9a.1)

THE FUMES given off by the soldering process can cause breathing difficulties, including asthma. However, these risks are normally associated with workers who solder all day, every day. Few amateurs are able to put in that amount of time, nevertheless it is important that the risk is kept as low as is reasonably possible, so good ventilation should be provided when soldering. This can be achieved by working near an open window or using a fan to blow the fumes away from you. You can buy special soldering iron attachments that will extract the fumes from right next to the tip of the iron. You need to assess how much soldering you are likely to do and take sensible precautions.

Eye protection (9a.1)

MOLTEN SOLDER can 'spit', and a shake of the soldering iron can send blobs of hot metal flying. It is therefore important that you get into the habit of cleaning excess solder from the bit after every joint and wearing safety spectacles or goggles to protect your eyes.

Goggles will also protect you from pieces of metal that might fly off as component leads are cut after being soldered.

And finally...

IF YOU ARE going to be soldering for the first time, you might like to try one of the many kits available on the amateur radio market. Most good radio kits include hints and tips on good soldering practice.

Good luck!

When you are not using a soldering iron, keep it on a stand.

Conductors & Insulators

Electron movement (3b.1)

SOME MATERIALS conduct electricity easily, that is to say that electrical current will flow through them with little resistance. These are known as *conductors*. Other materials do not conduct and are known as *insulators*. To understand why this is the case we need to consider the atomic structure of the different materials.

An atom is extremely small, and, as **Fig 1** shows, has a positively charged centre or nucleus. Orbiting around the nucleus are negatively charged electrons arranged in layers or 'shells', like the layers of an onion. The size of the positive

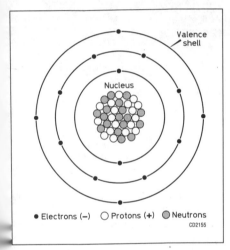

Fig 1: Diagram of an atom.

charge in the nucleus is balanced by the negatively-charged orbiting electrons.

The outermost electrons, being further away from the positive nucleus, are not so strongly attached to it. In conductors these outer electrons can be made to move from atom to atom by applying a voltage (sometimes described as electrical potential).

When we connect a battery to a simple DC circuit, the potential difference causes the electrons in the wire or bulb to move along the conductor. It is this movement of electrons, illustrated in **Fig 2**, that we refer to as *current*.

Good conductors & insulators (3b.2)

CONDUCTORS HAVE a very low resistance, because of their freely available mobile electrons. Metals are good conductors and amongst the most commonly used are copper, brass, steel, and nickel. Some of the best conductors, such as silver and gold, are too costly to use for most amateur radio purposes. Metal conductors are used for connecting pins, wires, coils, switches and many other current-carrying devices. You should note that some liquids and gases also conduct electricity - particularly if they contain impurities.

Insulators have electrons that are so tightly bound to their parent atoms that there are virtually no mobile electrons available. Therefore these materials conduct very little or no current.

Insulators, by definition, do not conduct electricity and are used to prevent current flowing where it is not wanted. Insulators are formed from a wide variety of substances, including dry wood, rubber, ceramics, glass, plastics, fibreglass and dry air. Some common insulators and their uses are:

- Ceramic and glass – high voltage insulators (e.g. at the end of wire aerials, because high RF voltages are present when transmitting)
- Rubber and plastic – to cover wires and cables
- Polythene and PTFE – to insulate the centre conductor of coaxial cable, coaxial plugs and sockets
- Plastics - to encase integrated circuits
- Fibreglass – the base material of many printed circuit boards

Words of warning!

A FLOW OF current *can* occur in an insulator if the potential difference between its ends is so large that it 'breaks down'. An insulator may also conduct if it is wet. Moist air will conduct electricity, and wood loses its insulating properties as its moisture content increases. This is extremely undesirable, as an insulator is intended to prevent current from flowing.

- If an insulator at the end of an antenna fails, it will allow the RF current to flow through it. The antenna will appear to have lengthened and the SWR will change, especially if any support ropes are wet.
- If the insulation on a mains cable breaks down a fatal electric shock could occur.

You should note that the human body is a conductor – after all, the signals from your brain to your muscles are electrical signals. This is one of the reasons why electricity is so dangerous. If the signals to your heart are 'swamped' by electricity from the mains, the organ will stop working and death will quickly follow.

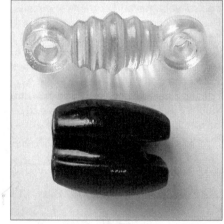

Typical antenna insulators made from glass and ceramic.

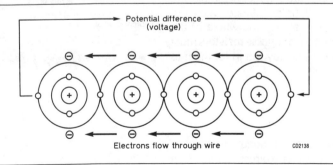

Fig 2: Electrons moving between atoms in a wire.

Components & Symbols

Recognising components (10a.1)

A WIDE RANGE OF electrical and electronic components are used in amateur radio equipment. You need to be able to recognise both the components themselves (for your construction project) and their symbols (as used in circuit diagrams).

Circuit symbols (3i.1)

ALL THE SYMBOLS used in the Intermediate syllabus are shown together in the table at the back of this book, but throughout the book each one is shown in the worksheet where it is first described. Some of those used in the Foundation training (the cell, the battery and the bulb) are not repeated in this book, but they are used in the worksheets. Let's have a look at a few new ones to start with.

Resistors

YOU SHOULD ALREADY know that resistance is measured in Ohms (Ω) and that values in Ohms (Ω), kilo-Ohms (kΩ), pronounced 'kill-ohms', and Mega-Ohms (MΩ), pronounced 'meg-ohms' are commonly encountered in amateur radio use. Circuit symbols are shown in **Fig 3**, but what are resistors?

As the name implies, they resist or oppose the flow of current. This allows us to control voltages and currents in circuits. Resistors can have fixed or variable values.

There are several types of resistor, but those you are most likely to come across will be 'carbon film'. These are made by coating a ceramic former with carbon, the value of the resistor being determined by how much carbon is deposited. Most fixed resistors look like small cylinders with connecting wires coming out of each end. Other types of fixed resistor include 'metal film' and 'wire wound'.

Variable resistors generally comprise a carbon track in the shape of a horse-shoe, with a sliding contact or 'wiper' to select how much of the track is in circuit. They may be 'pre-set' variables, which are adjusted using a screwdriver in a small slot and then left alone, or they can be truly variable with a control knob on a shaft, such as the volume control on a receiver.

Fixed resistors can be soldered into circuits either way round, but some variable resistors are logarithmic in nature. This means their resistance does not change evenly across the whole of the track. Logarithmic variable resistors must be fitted the right way round for them to work correctly.

Diodes

THE DIODE IS A component that allows current to flow in one direction only. This property makes it very useful in control circuits and in converting AC to DC, as you will see later in this book. As **Fig 4** shows, it also explains the arrowhead in its circuit symbol.

As the diode is a one-way device it must be fitted in circuit the right way round. In other words, we must observe the correct polarity. Most diodes are small cylindrical objects with two connecting leads, one at each end. A coloured band around one end represents the 'bar' across the arrow-head in the circuit symbol.

There are a number of special types of diode, some of which you will meet later in this book.

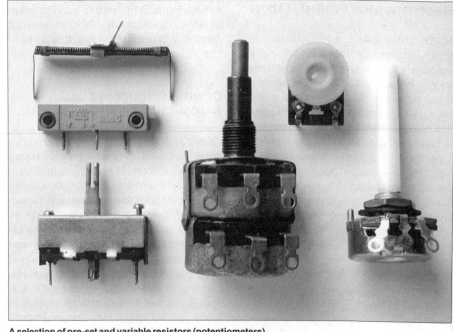

A selection of pre-set and variable resistors (potentiometers).

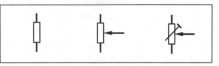

Fig 3: The circuit symbols for fixed, variable and pre-set resistors.

Fig 4: The circuit symbol for a diode.

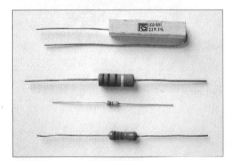

A selection of fixed value resistors.

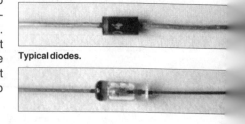

Typical diodes.

Transistors

THE TRANSISTOR IS an 'active' component that can be used as a switch, an oscillator or an amplifier. Again, you will see how later.

There are two main types of transistor – 'bipolar' and 'field effect'. Their circuit symbols are shown in **Fig 5**. They may look physically very similar, but they operate quite differently. Don't worry about the differences at this stage, just be aware that they have three connecting leads and must be fitted with each of the three leads in the correct position for them to work as intended.

The bipolar transistor has a collector, a base and an emitter, whereas the field effect transistor (FET) has a source, a gate, and a drain.

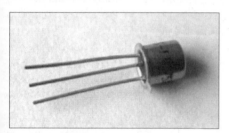

A typical transistor.

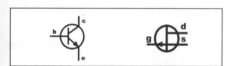

Fig 5: Circuit symbols for the (NPN) bipolar transistor and the FET.

A typical IC.

Integrated circuits

MANY HUNDREDS - if not thousands - of electronic components can be built into a device known as an integrated circuit, or IC. These are tiny silicon chips that are held in small plastic packages, with many connecting leads known as pins.

There are many different types of IC. Some have very specific uses, whilst others can be used in a variety of ways. Common amateur uses include (as **Fig 6** shows) audio amplifiers that require very few external components.

There is no single circuit symbol for the IC, but they are often depicted as a triangle, a square or a rectangle. It is important to make sure that an IC is fitted correctly, and generally they have an indent at one end and/or a spot next to pin 1 to help placement.

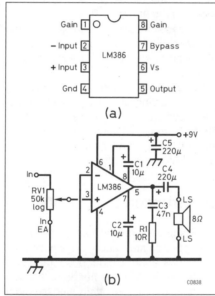

Fig 6: (a) Pin-out, and (b) circuit diagram of an audio amplifier based around an integrated circuit.

Exercise

TO BECOME MORE familiar with the component symbols covered so far, try drawing the circuits shown in **Fig 7**.

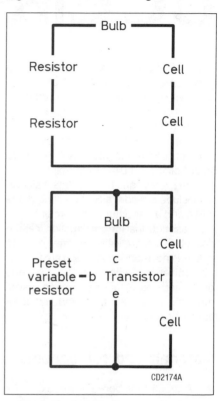

Fig 7: Re-draw these circuits, using the correct symbols.

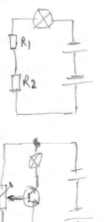

Building a Simple DC Circuit

THIS EXERCISE IS intended to help you to gain some soldering skills. Later exercises will use the circuit to help you learn more about the technical basics covered in the Foundation training.

The Intermediate syllabus requires you to demonstrate that you can make good solder joints and build a simple circuit using a battery, a bulb and two resistors. Later we will include a diode, a light emitting diode (LED) and a transistor in the circuit.

Please note that the construction method described below, which is very easy to follow, is not mandatory. If you are familiar with a different method (e.g. Vero-pins and perforated board) then by all means use it, but please maintain the same layout for ease of reference.

Materials and components

- A small test board - approximately 100mm square or larger, e.g. pin board, soft wood, balsa wood, non-silver side of a cake base, cork board
- Small bulb holder
- Bulb – 2.5 volt, 0.2 amp
- Resistor (R1) – 2.2 or 8.2Ω (colour code: red-red-gold or grey-red-gold) Note: The resistor colour code is explained on page 25.
- Resistor (R2) – 470Ω (colour code: yellow-violet-brown)
- Metal pins or small drawing pins – five required (must be brass or copper)
- Battery holder with 'snap' connector for two AA cells in series
- Battery connecting clip (preferably with red and black wires fitted as standard)
- AA cells – 2 required
- Connecting wire – 20cm of single stranded insulated wire
- Soldering equipment – soldering iron, stand, 20cm of multicore solder
- Tools – long-nosed pliers, wire strippers, wire cutters

Circuit design

IF YOU COMPLETED the circuit symbol exercises at the end of the previous worksheet, you will already have met the

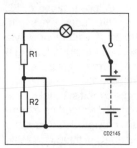

Fig 8: Circuit diagram of the simple DC circuit.

diagram for this simple DC circuit. It is shown again here in **Fig 8**. The physical layout of the circuit you are to build is shown in **Fig 9**.

Two resistors are wired in series with a bulb and battery, but the 470Ω resistor (R2) is shorted-out by a wire across it for the first tests.

The circuit is turned on by a 'shorting wire' (X), acting as a single-pole single-throw switch across points B and C in **Fig 9**. Current will flow from the battery, through the bulb and resistor R1. The bulb should glow modestly.

Construction (10d.2 & 10d.3)

1. Place pins on the test board in the approximate pattern shown, but not too far apart. Do not push them fully home. Ensure the heads of the pins are clean before attempting to solder to them (e.g. by rubbing them with wire wool or abrasive paper).

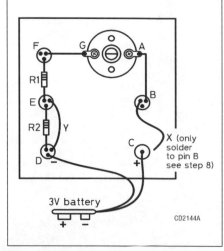

Fig 9: Layout of the simple DC circuit.

2. Fix the bulb holder to the board with double sided adhesive tape or suitable screws.

3. Strip insulation from both ends of four short lengths of wire (each about 3-4cm in length) and from the ends of the battery clip wires.

4. Solder the two resistors onto the pins in the positions shown. You may need to trap some of the leads under the pins to prevent the first resistor falling off as the second is soldered on.

5. **WARNING!** The pins get *very hot* and the solder can take a while to cool. Wait for it to do so, or you may burn yourself. *Do not* blow on the solder to cool it.

6. Solder a connecting wire between the top end of resistor R1 (pin F) and the bulb holder (point G). Make sure the joints are nice and shiny.

7. Solder a wire from the other side of the bulb holder (point A) to pin B.

8. Also at point B, connect one side of the shorting wire X.

9. Solder the red (positive) wire of the battery connecting clip to pin C.

10. Solder the black (negative) wire of the battery connecting clip to pin D.

11. Solder the shorting wire (Y) across R2 (pin E to pin D).

12. Check your connections, insert the batteries correctly and then turn the bulb on, by shorting B to C - do not solder the wire to pin C, just touch it with the bare wire.

13. Using the back of your finger just below the nail, feel the heat from the resistor R1 and the bulb. Even the battery will get slightly warm (consider why this heat is produced).

14. Once you are happy that the circuit works, switch it off, mark the board with the correct letters for each pin and put your name on it. You will be using the circuit in some later exercises.

Project Briefing & Tool Safety

The project (10e.1)

AS YOU ARE NO DOUBT aware, you are required to construct a simple radio-related project as part of your Intermediate Licence assessment. This is intended to ensure that you are capable of building your own transmitters - a key difference from the Foundation Licence. However, the project does not have to be a transmitter.

The project can be built from a kit purchased from one of the many suppliers, from an article published in a radio magazine such as *RadCom*, a technical handbook such as the *RSGB Radio Communication Handbook*, or from a circuit that you devised yourself (not that you are expected to have that degree of knowledge at this stage).

The choice of project is yours. An amateur band receiver would make an excellent project, but it is recognised that you may already have a perfectly good receiver as part of your M3 station. For that reason, other projects of similar complexity can be built to make a useful addition to your shack. The following list contains some examples of what would be acceptable:

- A crystal calibrator – useful for checking the frequency calibration of a receiver
- A dip oscillator – useful for checking the resonant frequency of tuned circuits
- An antenna tuning unit with built-in SWR indicator – an ATU on its own is considered too simple, but adding an SWR indicator requires a little more skill. An auto-ATU would also make an interesting, if challenging, project for a more experience student
- A Morse practice oscillator – very helpful if you intend to improve your CW
- An audio amplifier – perhaps to boost the AF output from a hand-held transceiver
- A QRP transmitter or transceiver – not to replace your existing rig, if you have one, but to have the satisfaction of achieving a contact with something you have built yourself

Some of the hand tools that you will find useful to have around your workshop or shack.

Whilst the choice is yours, you must check with your assessor before you commit to spending any money. The last thing you want is to turn up to the assessment to be told that all your hard work has been in vain!

Safety First!
The use of hand tools

BEFORE YOU START the project itself, the need to take care when carrying out any radio related construction must be stressed. Even simple hand tools can cause cuts and bruises, whilst power tools can be lethal in the wrong hands. However, by recognising the potential for harm and by following the following advice, you should minimise any risks involved.

Your tool kit

A GOOD TOOL KIT can be put together over a period of time, but there are a few essentials that you must have (or have access to). If you are taking part in a tutor-led course you may find everything is provided, but if you are learning alone you will need to put a small tool kit together. Those on tutor-led courses should also think about getting their own tools for after the course - they do come in handy!

You will need fewer tools for kit building than for building a project 'from scratch'. The amount of metalwork your project involves will determine whether you need things like hacksaws, drills and files, but the basic tools for kit building are:

- Soldering iron and stand
- Solder sucker and/or desolder braid
- Side cutters
- Wire strippers
- Sharp modelling knife
- Flat blade screwdriver (2-3mm)
- Cross point screwdriver (2-3mm)
- Small pair of pliers
- Multi-meter (see worksheet 8)
- A clear rule or tape measure
- A magnifying glass

Not exactly a 'tool' - but certainly another requirement - is somewhere to work. Not everyone has the luxury of a separate workshop and many radio amateurs successfully make use of 'the kitchen table'. If you fall into this category, please make sure you have a sheet of hardboard or similar material to protect the table from your construction activities!

Handling tools (9b.1)

SCREWDRIVERS, KNIVES, drills, saws and files all have sharp edges or points to do their intended work. If you don't take care, cuts and bruises will occur! Always pick up tools by their handles and keep your hands away from the 'sharp end' when you are using them. Look out for others who may be in the work area too, they may not be aware of what you are doing and are unlikely to appreciate the risks.

You should also remember that the leads of power tools can be a hazard, with a potential for tripping.

Using a vice (9b.2)

WHEN DRILLING, sawing or filing metal or printed circuit board materials, it is possible for the work piece to become a danger as it too can have sharp edges. It is important to ensure that whatever you are drilling or filing is securely fixed, in a vice for example, to prevent it slipping or spinning out of control. A small vice can be clamped to the aforementioned kitchen table... with the appropriate protection, of course.

Drilling hazards (9b.3, 9b.4, 9b.5 & 9b.6)

HAND DRILLS are less hazardous than powered drills, but it is still important to make sure that no part of your body is on the other side of the item being drilled and that the kitchen table is well protected!

In a power drill a chuck key is normally used to secure the drill bit. You must remember to remove the chuck key before you start drilling, or the key will become a missile as it is ejected at high speed!

A centre punch should be used to make a small indentation where you intend to drill a hole. The drill bit will be less likely to slip away when you start to drill. This is not only safer, it helps to make sure the hole is drilled exactly where you want it. Always start with a small 'pilot' hole, then use larger drills to enlarge the hole to the size you need. This is not only safer, but it makes a neater job.

When drilling sheet metal, the waste can form spiral shreds known as 'swarf' which is extremely sharp and can cause severe cuts. Gloves are not recommended, as they can loosen your grip on the tools and the work piece, so it is a matter of keeping your hands out of the way – use a vice.

Swarf can also be ejected from a drill bit at high speed, so eye protection such as safety spectacles or goggles are necessary. After drilling each hole, brush away the swarf using something like an old paintbrush. Use a dustpan and brush or vacuum cleaner to clear up after you have finished. *Never* try to blow swarf away, it can cause serious eye injuries!

A pillar drill is essentially an electric hand drill bolted to an upright stand. The drill is fixed for safety and a handle is used to lower the drill onto the work piece. This enables hands to be kept well away from the danger area, plus you have more control over the whole operation. Eye protection and care of the chuck key are still required though.

Further information

IF YOU ARE building a project for the first time, you would be well advised to seek guidance from a more experienced constructor. However, many of the kit suppliers include general construction advice with their kits and there is sound advice to be found in Appendix 1 of the RSGB book *Practical Transmitters for Novices*.

Fitting a 13 Amp Plug & Electrical Safety

Overhead power lines (9d.1)

IN MANY AREAS high voltage mains electricity is carried from sub-stations to houses on overhead power lines. Any contact with these power lines is likely to result in fatal injuries. You should also be aware that high voltages can arc across several metres, so don't even get close!

It is therefore extremely important that care is taken when using long ladders or installing tall antennas or masts. Make sure that you check both the area you are working in and its surroundings. If a long metal pole or ladder falls, it may strike power lines that are not directly overhead.

Pylons carry extremely high voltages and must be kept clear of.

Fuse rating (9d.2)

THE FUSE in a mains plug or power supply is a safety device. See **Fig 10** for the circuit symbol. A fuse is normally made from a short thin metal strip inside a glass or ceramic tube. The size of the metal strip is chosen so that it melts when a specified current flows through it. When the strip melts it breaks the circuit to stop the flow of current.

It is extremely important that the correct fuse is fitted to every piece of equipment in the amateur's home. If a fuse of too low a rating is fitted it could 'blow' when the equipment is working normally; if too high a rating is used it could continue to pass current even though a dangerous fault exists.

You will learn how to select the correct fuse in a later worksheet.

Fig 10: The circuit symbol for a fuse.

Residual current devices (9d.3)

WHILST FUSES provide good protection against too much current, they may not be good enough to prevent fatal electric shocks. You will remember that the human body is a conductor and it only takes about a 0.5A current flowing across a person's chest to stop their heart. A 5A fuse would therefore provide little protection if a fault only caused a current of 3A to flow.

A better safety precaution is a residual current device (RCD). These sense the mains AC and 'trip out' if any imbalance occurs between the live and the neutral circuits. Such an imbalance would occur if an item of equipment developed a leakage fault or you came into contact with a live conductor.

An RCD is designed to work at milliamp levels and provide a good margin of safety for the human body. Some houses have them fitted near the mains feed point, but you can also buy them built into plugs or extension leads. Every shack should have one!

Fitting a 13amp plug (10d.8)

AS YOU KNOW from your Foundation studies, mains cables normally have three wires, Earth (green and yellow striped), Live (brown) and Neutral (blue). As part of your Intermediate assessment you will have to fit a 13amp plug. The remainder of this worksheet explains how you do that.

To carry out this exercise you will need:

- A 13amp plug
- A short length of 3-core mains cable
- A set of wire strippers or a sharp knife
- A pair of side cutters
- Suitable screwdriver(s)

Take great care at each step of this exercise, as any errors could have serious consequences, *mains voltage can kill!* If any of the following steps go wrong, stop, cut off the end of the cable and start again. Also, before you start, check that

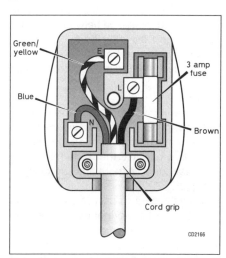

Fig 11: Wiring diagram for a 13amp plug, fitted with a 3amp fuse.

the piece of mains cable you use is *not* plugged into the mains! Refer to **Fig 11** throughout. A colour version of Fig 11 can be found on the inside back cover.

1. Remove about 40mm of the outer sheath of the cable. The best way to do this is with a cable stripper. If you use a sharp knife instead, run the blade around the cable about 40mm from the end, but don't cut so deep as to cut into the insulation of the three cores underneath.

2. If you used a knife instead of a cable stripper, in some cables the sheath can be removed by simply bending the cable back and forth at the point where you cut around it. If this does not happen you will need to make a light cut from the end of the cable to where you made the first cut. Again, be careful not to damage the inner cores.

3. Using a pair of side cutters grip the edge of the outer sheath and peel it back to the first cut then, using your fingers, pull off the excess.

4. Check for any damage to the inner cores. Bending the cable should expose any problems. If there are any nicks, stop and cut off the inner cores and start again. Only when you are satisfied that the inner cores are undamaged should you proceed.

5. Using the appropriate screwdriver, remove the top from the 13amp plug. Inside the plug, where the cable will sit, there is usually a short strip of fibre or plastic held in by two screws. This is known as the cord grip.

6. Loosen the two screws until you can pass the mains cable outer sheath through the space between the cord grip and the plug body. The outer sheath should be just visible on the inside of the cord grip, but don't tighten the cord grip screws yet.

7. Referring to Fig 11, find the terminal marked 'E' (Earth) and lay the green and yellow core through the plug so that it reaches the 'E' terminal in a neat curve, without any stretching.

8. Cut the core just beyond the terminal so it is a little longer than it needs to be.

9. Do the same with the blue core and the 'N' (Neutral) terminal.

10. Do the same once more with the brown core and the 'L' (Live, fused) terminal.

11. Remove the cable from the plug.

12. Using wire strippers, remove about 5mm of the insulation from the end of each of the three cores. If using the knife, use the same method as before but make sure that you do not damage the thin wires inside the insulation. If you damage any of the inner wires, stop, cut off all three cores and start all over again.

13. When you are sure that all three cores are OK, examine the plug top. Some plugs require you to thread the cable through a hole in the top, whilst others have a slot that allows you to fit the top after all the connections have been made. If yours needs to be threaded on, now is the time to do so. Make sure you thread it through in the right direction or you will need to take the whole thing apart at the end!

14. Re-fit the cable under the cord grip so that the outer insulation sheath is just visible on the inside again and make sure that the three cores are correctly positioned for their respective terminals.

15. Now tighten the two screws of the cord grip to hold the cable firmly in place.

16. Twist the inner wires of the green and yellow core together so that there are no stray stands.

17. Slacken the screw in the 'E' terminal, but do not remove it.

18. Push the twisted inner of the green and yellow core into the 'E' terminal and tighten the screw to hold it firmly in place.

19. Repeat the twisting and fixing of the blue and brown cores onto the 'N' and 'L' terminals respectively.

20. Check that the fuse is in place and not loose.

21. Finally, replace the plug top and tighten the holding screw.

Safety

DO NOT CONNECT the cable you have just fitted with a plug to the mains, unless it is connected to a piece of electrical equipment!

You may find that some pieces of equipment only have two inner cores in their mains cable. These should be the blue and brown. The plug should be fitted just as above, but without the earth connection. Leave the earth pin in place - it must not be removed! If the cable has three cores, all three must be used.

A 13amp plug is not always used with a 13amp fuse. Worksheet 12 explains how to select the correct fuse for various items of equipment.

Multi-meters & Units of Measurement

Introduction

ANY OPERATOR OF amateur radio equipment must be able to make some basic measurements to ensure that their equipment is working correctly.

Test equipment for amateur radio can be quite complex and expensive, but one of the most versatile and least expensive pieces of test equipment is the multi-meter.

Main features

A MULTI-METER usually has a large scale or digital display that shows the measurement and a range switch to select the type of test to be carried out.

There are two main types of multi-meter, analogue and digital. An analogue meter has a moving needle on a scale, whereas a digital meter displays a reading in numbers. Both types of meter have advantages and disadvantages, as you will find out shortly.

A multi-meter has two colour-coded test leads with probes on the ends. These probes are used to make contact with test points in the circuit you are measuring. You must *never* touch the metal probes whilst making a measurement. There may be high voltage or current present, so there is a safety issue. Also, your body could affect other readings. Read on!

Purpose of the multi-meter (3j.1)

YOU MAY NEED to take measurements to check components before you use them, test circuits as you build them, or find faults in existing equipment. Quite often, just checking the voltage at various points in a circuit can pinpoint where a component has failed.

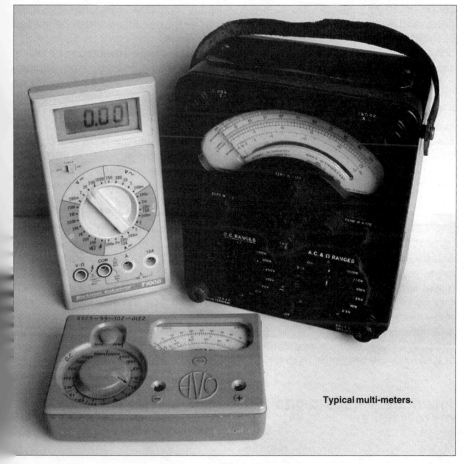

Typical multi-meters.

There are three main types of measurement carried out using a multi-meter:

- Voltage (potential difference)
- Current
- Resistance

Many modern meters have additional features such as transistor testers and frequency ranges. For Intermediate Licence purposes we will limit our coverage to the three main functions; voltage, current and resistance.

Units and abbreviations (3a.1)

YOU WILL REMEMBER from your Foundation training that the main units for measurements are Volts (V), Amps (A), and Ohms (Ω).

To make meters easier to read, each type of measurement is sub-divided into ranges. For example, the voltage measurement may be sub-divided into 10, 50, 250 and 1000V ranges. These figures are the maximum value that can be measured with that range selected.

You may be wondering why we have more than one range. Think about a 1V test signal. If we tried to measure it using the 1000V range it would hardly register, so a lower range would need to be used. If you have a meter and a 1.5V cell handy, try it!

A sensitive meter will have settings to measure very small values; milli-volts (mV) and milli-amps (mA). On the resistance scale you should find a range for much larger values, mega-ohms (MΩ). You will recall that 'milli' means 'one thousandth' and 'mega' means 'million'. A really sensitive meter may even have a micro-amp (μA) range, micro meaning 'one millionth', a tiny current.

The voltage and current ranges are also split between alternating current (AC) and direct current (DC). Therefore, as well as selecting the appropriate range, it is important to make sure you select the correct type of current, AC or DC. These may be selected by the range switch, or there may be a separate AC/DC switch. Some meters use different sockets for the test probes on some ranges.

Setting the multi-meter for use (3j.1)

TO MEASURE A potential difference, say across the terminals of a battery, you should first estimate the maximum potential difference you can expect to be present. For example, if you are measuring a DC circuit that is connected to a 12V supply, the maximum potential you would expect to find is 12V. You should then select the next highest range. In the example given previously, that would be 50V. The meter is then read, to measure the potential difference. If you find that the actual reading is very low, select a lower range and measure again. It is *always* better to select a higher range to start with, if you are at all unsure of what reading you will get.

On some meters you may have to move the test probes into a different pair of sockets to measure current (see photo on page 13). The rest of the procedure is similar to setting up for voltage measurements. However, it is not always as easy to estimate the maximum current you can expect to find, especially when fault finding.

The solution to this problem is to start with the highest current range and take a measurement to get a rough idea of what the current is. You can then switch to a lower range to take a more accurate reading.

Setting the meter to measure resistance is a bit more involved and we will come back to that in another worksheet.

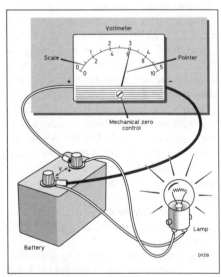

Fig 12: A voltmeter in parallel with a bulb.

Words of warning

USING A MULTI-METER with the wrong range selected can result in it being damaged beyond repair. It is *very* important to check your selection before making an actual test, and useful to return the meter to the 'off' position after testing.

Polarity (3j.3)

AS NOTED PREVIOUSLY, a meter has two test leads with probes on the ends. For voltage and current measurements involving DC, the test probes must be connected the right way round. This is known as 'observing correct polarity'.

To help with this the leads are usually colour coded, red for positive and black for negative.

Using the meter (3j.4)

WHEN YOU TAKE a voltage reading you are measuring the difference in electrical potential between two points. The test probes must therefore be placed across (i.e. 'in parallel with') the part of the circuit or component you are interested in (see **Fig 12**). Don't forget to double-check the range setting before use and observe the correct polarity.

For current readings you are measuring the flow of electricity through a circuit at a certain point. The meter must therefore be part of the circuit (i.e. 'in series with'). This generally means making a break in the circuit and attaching the test probes to each side of the gap (see **Fig 13**). Again, don't forget to double-check the range setting before use and observe the correct polarity.

If you are struggling to recall which test is which, just remember: 'volts across' the meter (like vaulting a fence) and 'current flows through' the meter (like water flows through a pipe).

You will have a chance to practise these skills in later worksheets.

Switching off

ALL DIGITAL METERS and many analogue meters use internal batteries to make them work. It is highly recommended that you get into the habit of switching your meter off after each test. Not only does this save the battery from discharging but it encourages you to check that you have selected the correct range each time.

Analogue versus digital meters (3j.2)

MANY NEWCOMERS assume that dig-

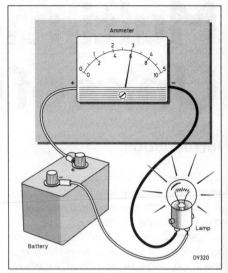

Fig 13: An ammeter in series with a bulb.

ital meters are better than analogue meters. In some cases this is true, but for other tests an analogue meter is a better instrument.

If you wish to take a straightforward measurement, say the voltage across a battery, then the digital meter is easiest to read. However, if you are making adjustments to a circuit and you are looking for a maximum value, or 'peak', the analogue meter is far better as the needle can be seen to change direction as the peak is passed. On the digital meter the display would be constantly changing and the maximum value can be missed quite easily.

The other main advantage of a digital meter is that it can be used with the probes applied the wrong way round. All that happens is that a 'minus' sign appears in the display, the value will be the same.

Ideally you should have one of each type of multi-meter, but if you are buying one for the first time it is worth noting that an analogue meter will do everything a digital meter will do *and* be able to show you 'peak' readings. You would be well advised to buy the best meter that you can afford, as cheaper models are less robust and much less accurate.

More information

IF YOU ARE interested in building-up your range of test equipment beyond the basics, the RSGB book *Test Equipment For The Radio Amateur* contains details of how to use the different pieces of equipment together with many designs to build.

Measuring Potential Difference

YOU WILL RECALL that potential difference is the force that causes negatively charged electrons to flow around a circuit and that the flow of electrons is called current. Potential difference is measured in Volts and you may find that potential difference is often referred to as 'voltage'.

In the simple DC circuit you have built, the potential difference is provided by the two 1.5V cells in series, giving a total of 3V. There is one resistor R1 (2.2 or 8.2Ω) in series with the light bulb. Note that the light bulb also acts like a resistor, but it glows white hot, giving out light. The resistor(s) cause the potential difference, to fall as the current flows through them.

How to measure potential difference (3j.4 & 10d.4)

TO MEASURE potential differences we need to use a multi-meter set to work as a 'voltmeter' and it needs to be placed *across,* or 'in parallel with', the component being measured. **Fig 14** shows an example. The voltmeter has a 'high internal resistance', which means it hardly affects the circuit at all.

To put this into practice, turn on the simple DC circuit you constructed earlier by shorting out the points B and C with the wire (X) and follow the instructions below.

1. Set the multi-meter range switch to DC volts and select the next highest range to the expected 3V (the exact range will depend on your meter).

2. Measure and note down the potential difference across the bulb, by placing the red (+) lead on the positive side of the bulb at pin A and the black (-) probe on the negative side of the bulb at pin G. (Note: If using an analogue meter, ensure that the red probe is on the positive side of the bulb, otherwise the pointer will try to move backwards which could cause damage.)

3. Measure and note down the potential difference across R1, again observing correct polarity with the meter probes.

4. Now do the same for the potential difference across the battery at points D and C.

5. Turn off the circuit, by disconnecting the wire between point B and C.

6. Finally, measure and note down the potential difference across points B and C whilst the circuit is off.

7. Switch the meter to its 'off' position.

Results

IF YOU ADD THE potential difference across the bulb to the potential difference across the resistor R1, it should equal the potential difference across the battery when the circuit is turned on. If it doesn't quite add up, it is probably because the readings

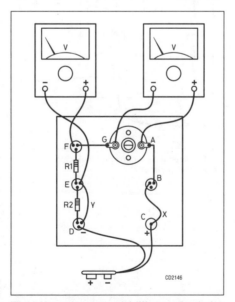

Fig 14: Measuring a potential difference, by placing meters in parallel with components.

were difficult to measure precisely.

In a series circuit like this one the current flows through each component in turn. The potential difference across the battery is divided between the components in the circuit. The potential difference across each of those components depends on its resistance but if you add them all together it should be the same as the value across the battery.

The potential difference across the points B and C when the circuit is turned off may be slightly higher than that across the battery when the circuit is switched on. This is because the potential difference across the battery drops slightly when current is flowing through the circuit.

Summary

THE KEY POINTS to note from this exercise are:

- Potential difference causes current to flow.
- Potential difference is measured across (or 'in parallel with') components.
- In a series circuit, the potential difference supplied by the battery will always be divided between the components in the circuit.
- A voltmeter has its own very high internal resistance, so it only draws a tiny amount of current in making the meter operate. This means that it doesn't significantly alter the actual current (and therefore the potential differences) in the circuit that it is trying to measure.

Measuring Current

Units of measurement (3a.1)

AS WE LEARNED in worksheet 3, current is the flow of electrons around a circuit. It is measured in amps (A), or sub-divisions of an amp, i.e. milli-amps, mA (thousandths of an amp) or micro-amps, µA (millionths of an amp).

How to measure current (3j.4, 3j.5 & 10d.4)

TO MEASURE CURRENT we need to set our multi-meter to work as an 'ammeter'. **Fig 15** shows a meter being used to measure current.

A multimeter is in its most sensitive and most vulnerable state when it is being used as an ammeter. An analogue meter in particular is most at risk of damage when measuring current. If you ignore this point, one day you may experience the unfortunate situation of seeing wisps of smoke rising behind the meter glass! It is therefore important to use the correct or *safe* current range and to have the meter probes the correct way around.

Current is measured in *series* with the circuit being tested. This means that *all* the current flowing through the circuit being tested has to pass through the ammeter.

In a series circuit it will measure the same at all parts of the circuit. If the meter is taken out of the circuit during the measurement, no current will flow as the circuit will have been broken.

As we did with measuring voltage, follow the instructions below to see how this works in practice using the simple DC circuit you built earlier.

1. The circuit should be 'off', i.e. points B and C should be open.

2. Set the multi-meter to DC amps and onto a high scale (e.g. 10A). Remember, you may have to change the test leads into different sockets.

3. Double check the range setting and then place the red (+) probe on to pin C and the black (-) probe onto pin B.

4. Current will flow through the meter and the bulb should light, but on this range the meter should not move much because the current is quite small. Can you read how small?

5. Once you have confirmed that only a small current is flowing, switch to a lower range, say 1A maximum. If the meter still reads well under half scale, switch to a lower range still, say 500mA. Remember, 500mA is half an Amp, or 0.5A.

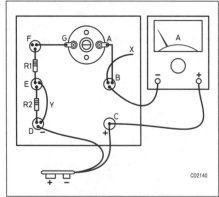

Fig 15: Measuring a current, by placing a meter in series with a circuit.

6. Note down the meter reading for later use, then switch the meter to its 'off' position.

If you are in any doubt that the same current is flowing through all parts of the series circuit, try breaking the circuit between the bulb and resistor, shorting B to C and measure the current between pin F and the bulb contact G.

Summary

THE KEY POINTS to remember from this exercise are:

- Current flows *through* complete circuits.
- Current is measured in *series* with the circuit.
- You can measure the current anywhere in a series circuit and it will be the same.
- An ammeter has a very low series resistance, so it doesn't alter the current that it is trying to measure.

Remember: A multi-meter is at its most vulnerable when measuring current. This is why you should always start with the highest current range, before switching to a lower range. Never leave a multi-meter set to a DC current range when not in use. Most meters have an 'off' position, but if yours doesn't you should leave it on a high AC volts range in case a voltage is applied accidentally.

Licence Conditions I

IN YOUR FOUNDATION Licence training you learnt a little about the rules you needed to follow as a Foundation Licence holder. At the Intermediate level you will need a better understanding of the regulations set out in the 'Terms, Provisions and Limitations Booklet BR68/I', particularly those covering the additional privileges you will gain at the Intermediate level.

Although you must still abide by the terms of your Foundation Licence for now, you should put BR68/F away and start using BR68/I for your studies. Doubtless you will be pleased to know you do not need to remember all of it! Much of the booklet covers situations that will not apply to you, certainly at first. If you take up other aspects of amateur radio later, you can re-read the booklet to see what rules apply.

In this worksheet, and a later one covering other parts of the Licence conditions, the style will be to give the syllabus item **in bold**, the actual text from BR68/I (i.e. the official rules) *in italics*, and some words of explanation in plain text.

The extracts from BR68/I are correct at the time of printing, but you should ensure you have the latest copy for study. This can be checked by looking at the OfCom web site (www.ofcom.org.uk) or by obtaining a current copy of BR68/I from the OfCom publications office.

1a.1 Recall the various types of amateur licence. Foundation, Intermediate, Full and recognise their callsigns

You already know that there are 3 levels of licence and that they each have different callsign series:

- Foundation: M3QQQ
- Intermediate: 2E0QQQ
- Full: M0QQQ

The same Regional Secondary Identifiers (shown in **Fig 16**) are used at the Intermediate level, depending on your location:

- D - Isle of Man
- E - England
- I - Northern Ireland
- J – Jersey
- M - Scotland
- U – Guernsey
- W - Wales

You will recall that for Foundation and Full callsigns these secondary letters are inserted between the M and the number in the callsign prefix (e.g. MW for Wales, MU for Guernsey, etc.). However, for the Intermediate Licence callsigns, E is used for England and is replaced by the Secondary Regional Identifiers for the other parts of the UK. So the Intermediate Licence holder, 2E0QQQ, would be an amateur based in England. Had the station been registered in Wales the callsign would have been 2W0QQQ.

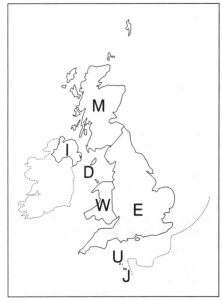

Fig 16: Regional identifiers within the UK.

Historically (but not for the exam), UK amateur callsigns began with a G. However, the G series filled up and the M series was started a few years ago. You may also come across GB callsigns which are issued to special event stations and repeaters.

2a.1 Recall that only the licensee personally may use the station

2a.2 Recall that an Intermediate licensee may operate the Station of a Full licensee, using the call sign of and under the direct supervision of the Full licensee

2(8) The Station shall be operated only by the Licensee personally.

AS WITH THE Foundation Licence, only you may use your station. As an Intermediate licensee it is assumed that you do not

The front cover of BR68/I, the Terms and Limitations booklet for the Intermediate Licence.

have enough experience of amateur radio to supervise others using your station (whether they are licensed themselves or not).

If two amateurs live at the same address and share equipment, the 'Station' is identified by the callsign being used at the time. Each person must use their own callsign and use the equipment as if it were their own. Similarly, if another licensed amateur visits your shack and uses your equipment, they must use their own callsign (see syllabus item 2a.1) and enter the details in their own logbook. Remember: if *your* callsign is being used, only *you* can be on the microphone.

On the other hand, if you visit a Full licensee and use his or her station 'under supervision', you may use his or her callsign. You should also know that 'under supervision' means that the Full Licencee is in the room with you and able to help if you are doing anything outside the licence conditions.

In such circumstances, you may use all the privileges of the Full licence, i.e. transmit power, bands and modes of transmission. Of course, if you want a QSL card for a contact you will have to use your callsign, enter the call in your logbook and limit yourself to the conditions of your licence.

2b.1 Recall that messages must be limited to matters relating to technical investigations and remarks of a personal nature

2b.2 Recall the Station must not be used for business or advertising purposes

1(2) The Licensee shall address Messages only to other licensed amateurs or the stations of licensed amateurs and shall send only:
(a) Messages relating to technical investigations or remarks of a personal character; or
(b) Signals (not enciphered) which form part of, or relate to, the transmission of Messages.
3(3) Except as provided by sub-clause 3(4), the Licensee shall:
(a) have no pecuniary interest (direct or indirect) in any operation conducted under this licence; and
(b) Except in the case of activities on behalf of a non-profit organisation established for the furtherance of amateur radio, not use the Station for business, advertisement or propaganda purposes including (without limiting the generality of the foregoing) the sending of news or messages of, or on behalf of, any social, political, religious or commercial organisation.

As for the Foundation licence, you should only speak to other amateur stations and talk about personal topics and your amateur radio activities. You should also remember from your Foundation studies that you must not carry out any business activities using amateur radio or pass messages for other groups that are not amateur radio related.

This means that it would be quite acceptable to advertise forthcoming radio club meetings to other amateurs you are in contact with, but it would not be acceptable for an amateur radio shop to advertise what is on special offer.

2b.3 Recall that messages may be used to aid User Services and who they are

10(1)(1) "User Service" means the British Red Cross Society, the St John Ambulance Brigade, the St Andrew Ambulance Association, the Women's Royal Voluntary Service, the Salvation Army, HM Coastguard, the Chief Emergency Planning Officer or any United Kingdom police force, fire or ambulance service, health authority, government department or utility services.

As a service to the community, Intermediate and Full Licence holders are permitted to pass messages on behalf of the 'User Services'. They are the organisations listed above. Many of those bodies will have their own communications, but often communication between different organisations is not readily available. Some bodies such as the Women's Royal Voluntary Service do not normally need communications and do not have their own. However, at a major event such as a county show, or in an emergency, they may have need of good radio communications.

The Intermediate Licence allows you to provide such a service, provided the User Service has asked for assistance (in writing). Amateur interest groups such as the Community Radio Volunteer Service and RAYNET are set up specifically for such activities. If you are interested in joining one of these groups, your Local Authority Emergency Planning Department should be able to put you in touch with one.

The list of User Services includes 'utility services'. These are generally taken to be your local gas, electricity and water supply companies.

2c.1 Recall that the licensee must transmit the call sign printed on the Validation Document during CQ calls, at the start and finish of all periods of transmission and every 15 minutes during long periods of transmission

7(1) Subject to sub-clause (1A) below, which does not apply to operation via repeaters during transmissions, the Licensee shall transmit the call sign specified in the Validation Document:
(a) during initial calls ("CQ" calls);
(b) at the beginning and at the end of each period of communication with a licensed amateur and when the period of communication is longer than 15 minutes, at the end of each interval of 15 minutes;
(c) at the beginning of transmission on a new frequency (whenever the frequency of transmission is changed);
(d) by the same type of transmission that is being used for the communication;
(e) on the same carrier frequency that is being used for the communication.

(a) simply says that you must give your callsign when you make a 'CQ' call – inviting anybody who hears you to join in a conversation.

(b) means that you must give your callsign when you make contact with another station and again when you finally break contact. Also, if your conversation lasts more than 15 minutes, you must give your callsign at not more than 15 minute intervals. In practice amateurs give their callsign rather more frequently, but often not on every 'over'.

(c) makes the rather obvious point that if you change frequency you will need to give your callsign again.

(d) says that if you are using a particular type of transmission (e.g. single sideband) then you give your callsign in the same type of transmission, but see syllabus item 2c.2, below.

(e) goes hand in hand with (c), requiring you to give your callsign on the same frequency as the one you are using for your contact.

2c.2 Recall that the callsign must be sent in either voice or Morse code at least every 30 minutes during periods of transmission using other modes

7(1)(f) by Morse telegraphy or telephony, at the end of each 30 minute period during which transmissions are sent from the Station (unless already transmitting in Morse telegraphy or telephony).

Put simply, this means that if you are not using speech or Morse, you need to give your callsign every 30 minutes using speech or Morse so that somebody who does not have suitable equipment for your type of transmission can still understand the callsign.

For example, PSK31 is a popular computer-based mode of transmission that can only be decoded using appropriate software. It is therefore important that you transmit your callsign in Morse or by voice at least every 30 minutes, so that those without PSK31 facilities can tell who is transmitting. Most good software can be configured to do this for you.

You should note that even on a short transmission you should give your callsign in speech or Morse at least once.

2c.3 Recall the meaning of "Main Station Address", "Temporary Location" and "Mobile"

1(7) "Station" means the station of the licensee at the Main Station Address, a Temporary Location or while Mobile, as the case may be.
1(8) The licensee shall operate the Station only:
(a) at the main Station Address ("Main Station Address" means the main station address of the licensee set forth in the Validation Document);
(b) at a Temporary Location ("Temporary Location" means a location, other than the Main Station Address, in the United Kingdom, and in any fixed position);
(c) while Mobile ("Mobile" means located in the United Kingdom in any vehicle, as a pedestrian or on any vessel in Inland Waters).

This should be reasonably clear. The Main Station Address is the one you notified when you applied for the licence. In normal circumstances this will be your home address.

A temporary location is anywhere else when you are not on the move. This could be a friend's house, a hotel or a campsite. Operating in a contest is frequently done away from home. 'Field Days' require you to be independent of mains supplies, for example. Such operation should be logged in the normal way, together with the location of the Station. You must also say where you are (see syllabus item 2c.4).

Mobile means out and about, and includes short periods of stationary operation such as sitting on a park bench or waiting in the car whilst somebody goes shopping. There is no requirement to say where you are, although many amateurs do, and there is no need to log transmissions.

2c.4 Recall that the licensee must give the location of the Station, to within 5km, at least every 30 minutes when operating from a Temporary Location

7(2) At a Temporary Location, the Licensee shall:
(a) use the suffix "/P" with his callsign and give the location of the Station every 30 minutes to an accuracy of at least 5km by a generally used identifier.

When you are at a temporary location, an Intermediate callsign becomes something like 2E0QQQ/P, which is normally spoken as "2E0QQQ stroke P". You also need to say where you are, so that somebody following your instructions will be able to get within 5km of you. Note (v) to the licence conditions booklet gives some examples of what can be used. You will need them when out operating, but not for the exam.

2c.5 Recall that the appropriate secondary regional locator (D, I, J, M, U, W, E) must be used when operating away from the Main Station Address

7(4) When away from the Main Station Address, the Licensee shall use the appropriate Regional Secondary Locator specified in note (w) to this Booklet.

The list of Regional Secondary Locators was shown earlier in this worksheet. Consequently, if 2E0QQQ was temporarily located in Wales, the callsign would become 2W0QQQ/P or, if mobile, 2W0QQQ/M.

Of course, if your main station address is in a part of the UK other than England your callsign will include the appropriate Regional Secondary Locator. It follows that if the Scottish station, 2M0QQQ, went on holiday to Guernsey, the callsign from a hotel in Guernsey would be 2U0QQQ/P.

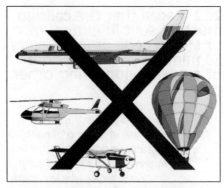

Operating an amateur radio transmitter from any airborne vehicle is forbidden.

2c.6 Recall that the licence does not permit operation from an aircraft or vessel except on inland waters

You may be familiar with the request to switch off all mobile phones when on an aircraft. This is due to the possibility of interference with the aircraft's own communications systems. The Intermediate Licence prohibits airborne operation for the same reasons. Operation at sea or in tidal waters is also prohibited at Intermediate licence level, but you may use your station mobile (/M) from a canal boat or other craft on inland waterways.

Operating an amateur radio transmitter from a boat or ship is permitted on inland waterways, but not on tidal rivers or at sea.

2c.7 Recall that the Intermediate Licence does not permit operation outside the UK

At the Full licence level the qualification requirements throughout Europe and many other countries are harmonised. The Harmonised Amateur Radio Examination Certificate (HAREC) is available to all Full European licence holders, simplifying travel with amateur radio. However, not all countries have Foundation, Intermediate or Novice licences, and where they do their standards are not coordinated. Consequently, other countries do not generally recognise the UK Foundation or Intermediate licences for use abroad. This is also the reason why operation on tidal waters or at sea is limited to Full licensees. However, it is worth checking with the authorities in the country you are travelling to before you go.

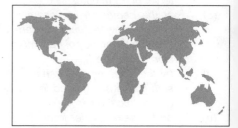

Its a big old world, but to operate your transmitter outside of the UK & Northern Ireland you will have to progress onto a Full Licence.

What Next?

THAT COMPLETES all you need to know about who can operate your station, what messages you can send and how to identify yourself. It looks like quite a lot of work, but it should not take long for it to become reasonably familiar. It is the sort of material that is best read, discussed as a topic with your instructor or another amateur, then put down while you do something else. When you have a few spare moments take time to read this worksheet and the second one on Licence Conditions again.

In the second worksheet we will look at the rules and regulations that apply to some other aspects of amateur radio such as beacons, digital operations, keeping your Log on a computer, and Electromagnetic Compatibility (EMC).

Calculating Input Power

POWER IS A measure of the work done by an electrical circuit, or by any device, in performing its task. Work is normally done by converting one form of energy into another, but we tend to refer to the energy being consumed or used. For example, in a kettle the electrical energy is transformed into heat to boil the water. The energy is provided by the power supply. This could be a battery, the mains, solar cells, etc.

The rate at which energy is used by an electrical system is called the input power, i.e. the power going into the system. The 'system' could be a single device such as a bulb, or a piece of equipment with many circuits, an amateur transmitter perhaps.

Your Foundation training taught you that power (W), measured in Watts (W) is equal to the potential difference (V), measured in Volts (V), multiplied by the current (I), measured in Amps (A). The Watt is named after James Watt who did much work on improving the early steam engine. The equation W = V x I can be rearranged so that you can calculate any one of the values, provided you know the other two.

The triangle shown in **Fig 17** is an easy way of remembering the appropriate equation. Cover with your finger the symbol you need to work out, and the two remaining tell you how to use them; side by side means multiply, one over the other means divide. If you are in any doubt about which letter goes where, ask yourself "*What* is at the top?" and you have the answer; *Watt* is at the top!

It is important that the correct units of measurement are used in this equation; Volts, Amps and Watts. Make sure you convert any other values before you make any calculations otherwise you will get the wrong answer. To convert from millivolts (mV) to Volts (V) you move the decimal point three places to the left. In other words, 1000mV = 1V. To convert from milliamps to Amps move the decimal point three places to the left. In other words, 100mA = 0.1A.

Using the test circuit (3j.6)

SO MUCH FOR the theory. How does it work in practice?

To calculate the input power used by the light bulb in your circuit, you need to measure the current flowing through the bulb and multiply it by the potential difference across the bulb. You have already meas-

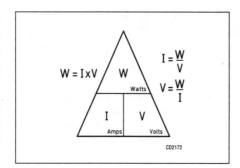

Fig 17: Cover the letter of what you want to know and the remaining letters show you how to arrange the formula.

ured these values in an earlier exercise, but if you wish to repeat it here, carry out the following:

1. Turn on the circuit by shorting out the switch at points B and C. The bulb will light.

2. Place the voltmeter on a suitable low range (e.g. 5V DC).

3. Measure the potential difference across the bulb.

4. Turn the circuit off by opening the switch at points B and C.

5. Set the ammeter to a suitable range (e.g. 1A or 500mA).

6. Check the polarity and measure the current flowing through the whole circuit (the bulb will light when you connect the meter).

7. Multiply the current (in Amps) by the potential difference (in Volts). If necessary, convert the units first. The result will be the input power to the bulb in Watts. You should find the power to be somewhere in the region of 0.25W.

It is worth remembering that the bulb is converting the electrical energy into both light and heat. The light and the heat are in fact the output of the bulb, in the same way that radio waves are the output of transmitters.

The other components in the circuit also convert energy to some extent, for example the resistor and battery will become slightly warm. We have only measured the potential difference across the bulb, so the input power we calculated is for the bulb alone, not the entire circuit.

Practical example (9d.2)

YOU CAN CALCULATE which fuse to fit using the equation for input power. Fuses are rated in Amps, electrical equipment is normally labelled with its power consumption (input power) in Watts and we know that the mains supply operates at 230 Volts. If we know the input power and the potential difference, we can calculate the current:

Amps = Watts / Volts

As an example, an amateur's station mains power supply unit is labelled as consuming 500 watts of power, so...

500 / 230 = 2.17A

Mains fuses come in 3, 5 and 13 amps, so you should fit the lowest fuse with a rating higher than the current required. In this case a 3 amp fuse would be suitable.

A note about output (3j.6)

CALCULATING THE input power is one way of 'rating' a transmitter, but the Licence conditions refer to "RF power supplied to the antenna". This means that we must have a method for measuring *output power*. In reality this is quite simple using an RF power meter. Many are available on the amateur radio market. Some double-up as SWR indicators. Other designs incorporate a dummy load and can be very useful pieces of test equipment. The author built one from a design that was published in the August 1984 edition of *Practical Wireless* and it is still in use.

It is perhaps worth noting that input power is *always* greater than output power. In many cases output can be as little as 50% of the input.

Summary

THE KEY POINTS to remember from this worksheet are:

- Input power is the energy supplied to the circuit.
- Input power of the bulb is the rate at which energy is supplied to the bulb.
- Output power is always less than input power
- Power (Watts) = Volts x Amps

Operating Practices & Procedures

IF YOU HAVE BEEN on the air with your Foundation Licence, you will know that enthusiasts from all over the world share the amateur radio bands. You will also have realised that it can be difficult to find enough space for everyone who wants to use the airwaves at the same time. Over the years, radio amateurs have developed procedures to help alleviate these problems.

Abbreviations and Q-codes (8a.1 & 8b.1)

MANY AMATEUR RADIO abbreviations were originally developed for use with Morse code. Given that Morse is a relatively slow method of communication, operators standardised on a series of abbreviations for commonly used messages, to speed up their routine transmissions.

These days some of these abbreviations are used with speech and data as well as Morse code, so even amateurs who have little understanding of English are able to exchange basic information with each other. There is no real need to use them between amateurs who understand each other's language, and it is better to use plain language.

The first group are known as Q-codes, because of their first letter. They have no obvious logic and have to be learned by heart (if you don't know them already). There are many Q-codes in use, but for the purposes of the Intermediate exam you only need to know those shown in **Table 1**.

It is quite common to use some of the Q-codes as questions. For example, 'QRL?' can mean 'Is this frequency in use?' and 'QSL?' can be a short way of saying 'Have you received my information correctly?'

Next we have a few abbreviations (see **Table 2**). Some are only found in Morse, but you may also see them used in data (keyboard-to-keyboard) conversations.

RST reports (8c.1)

RADIO AMATEURS are interested in knowing how well their signals are being received. One of the first items of information that will be exchanged during a contact is a signal report.

With Morse, amateurs use the three-number RST code for signal reports. The digits specify characteristics of the received signal, and stand for Readability, Strength and Tone. When using voice, Tone is omitted. You should recall giving a 'Readability and Signal' (RS) report as part of your Foundation assessment. You now need to remember some of the codes.

Readability is specified on a scale of 1 to 5, where 1 means an unreadable signal, and 5 means perfectly readable. You do not need to remember the numbers in between 1 and 5 for the exam.

Strength is specified on a scale of 1 to 9, where 1 means faint signals which are barely perceptible and 9 means extremely strong. You do not need to remember the numbers in between 1 and 9 for the exam.

Tone is specified on a scale of 1 to 9, where 1 is an extremely rough hissing note and 9 is a pure note. You do not need to remember the numbers in between 1 and 9 for the exam.

Putting this into practice, a signal that is perfectly readable and very strong would be described (on voice) as RS 59 or, more often, as 'five and nine'. Be aware of the distinction between readability and strength. A signal could very strong but so distorted that you cannot understand a single word, in which case the correct report would be Readability 1 Strength 9, or 'one and nine'. On the other hand, you might give a report of Readability 5 Strength 1 ('five and one') for a signal that was very faint but perfectly understandable.

As far as the Tone designation is concerned, it would be very unusual these days to hear anything other than a T9 report. In years gone by it was harder to generate a good Morse note and some very poor signals could be heard. Nowadays, even the simplest home-built equipment should create a good quality tone.

Prefixes (8f.3)

THE ABILITY TO contact stations in different countries is part of the pleasure of being a radio amateur; indeed you may already have some foreign contacts in your Foundation logbook. Gradually you will learn the different callsign prefixes used by stations around the world. At Intermediate level you are expected to know some of the common ones that you are likely to come across.

So, if you contacted a station with the callsign W3LPL, you would know its location was the United States. There are countries, including some of those in **Table 3**, which have several prefixes allocated. You will become familiar with these over time, but you only need to remember those listed in the table for the exam.

CQ	General call (any station may reply)
DX	Long distance (on HF this usually means outside your own continent)
SIG	Signals
UR	Your
WX	Weather
DE	From (example, 2E0QAC DE M0QTC)
K	Go ahead (your turn to transmit)
R	Roger (that is, transmission received and understood)

Table 2: Commonly used abbreviations.

QRL	The frequency is in use
QRM	Interference from other stations (M = man-made)
QRN	Interference from static/thunderstorms (N = natural)
QRP	Low power
QRT	Closing down my station
QRZ	Who is calling me? (Who'Z calling?)
QSB	Fading, usually signals going up and down in strength
QSL	Transmission successfully received (as in QSL card)
QSO	Contact with a station
QSY	Change frequency
QTH	Location, usually the nearest town (H = home)

Table 1: Commonly used Q codes and their meanings.

EI	Irish Republic, Eire
F	France
I	Italy
JA	Japan
PA	Netherlands
VE	Canada
VK	Australia
W	USA
ZL	New Zealand

Table 3: Some common callsign prefixes and their associated countries.

QSL cards (8f.2)

SOME RADIO AMATEURS like to collect QSL cards, which are postcards that confirm that a contact has taken place. As well as a block of standard information, they may also show pictures such as the station being used or a local landmark. Some QSL cards are completely plain, whilst some are extremely elaborate.

Radio amateurs who are keen to receive confirmation of a contact, perhaps with a rare country, will send a QSL card direct to the operator concerned and request a reply. To reduce expenses, many national societies operate 'QSL Bureaux', which collect batches of cards from their members. These are sorted and then sent to corresponding bureaux in other countries who distribute them in turn to their membership. The RSGB operate the UK QSL bureau. Some clubs, such as the G-QRP Club, operate their own bureaux, but these are only available to their members.

To receive cards from a national QSL bureau you need to send stamped, addressed envelopes to your nominated QSL manager. The envelopes are returned to you when they are full. Of course, this procedure can take several months, sometimes years, but the costs are lower than sending individual cards direct. You should note that anyone can receive cards through the RSGB QSL Bureau (see **Fig 18** overleaf for a diagram of how it operates), but only RSGB members can send cards through it.

Awards and contests (8f.4)

QSL CARDS ARE often required for the awards and certificates that national organisations issue to radio amateurs. These recognise achievements such as contacting a certain number of countries, callsign prefixes or islands, or for achieving long distance contacts. Operators who are interested in this aspect of the hobby may be said to be interested in 'DX', using the term loosely to mean any wanted station (not necessarily a long distance away).

Trying to find DX stations before anyone else is one part of the competitive nature of amateur radio. Another is participation in contests. These take varying forms, but usually involve contacting as many stations as possible in a given time. There is often an element of passing information accurately (e.g. callsign, contact serial number, location), which is checked after the event. Rules for each contest vary and are generally published by national societies, who may also collect and adjudicate the results.

Further information (8f.1)

MANY ORGANISATIONS publish 'Call Books' which contain lists of international prefixes, and it is useful to have this information available when you are operating. Call Books also list radio amateurs in their own country, often with their names and addresses.

The *RSGB Yearbook* is published every year. As well as UK and Irish callsign listings, it contains a prefix list and lots of other useful information on operating practices, QSL information, awards and contests. For those with access to the Internet, there are also several websites that contain international callsign listings.

The RSGB also publish an *Operating Manual* that covers many aspects of the hobby in some detail.

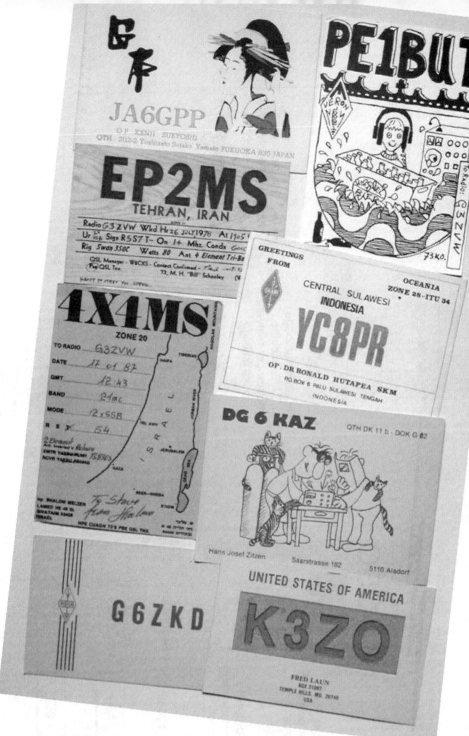

A selection of QSL cards from around the world.

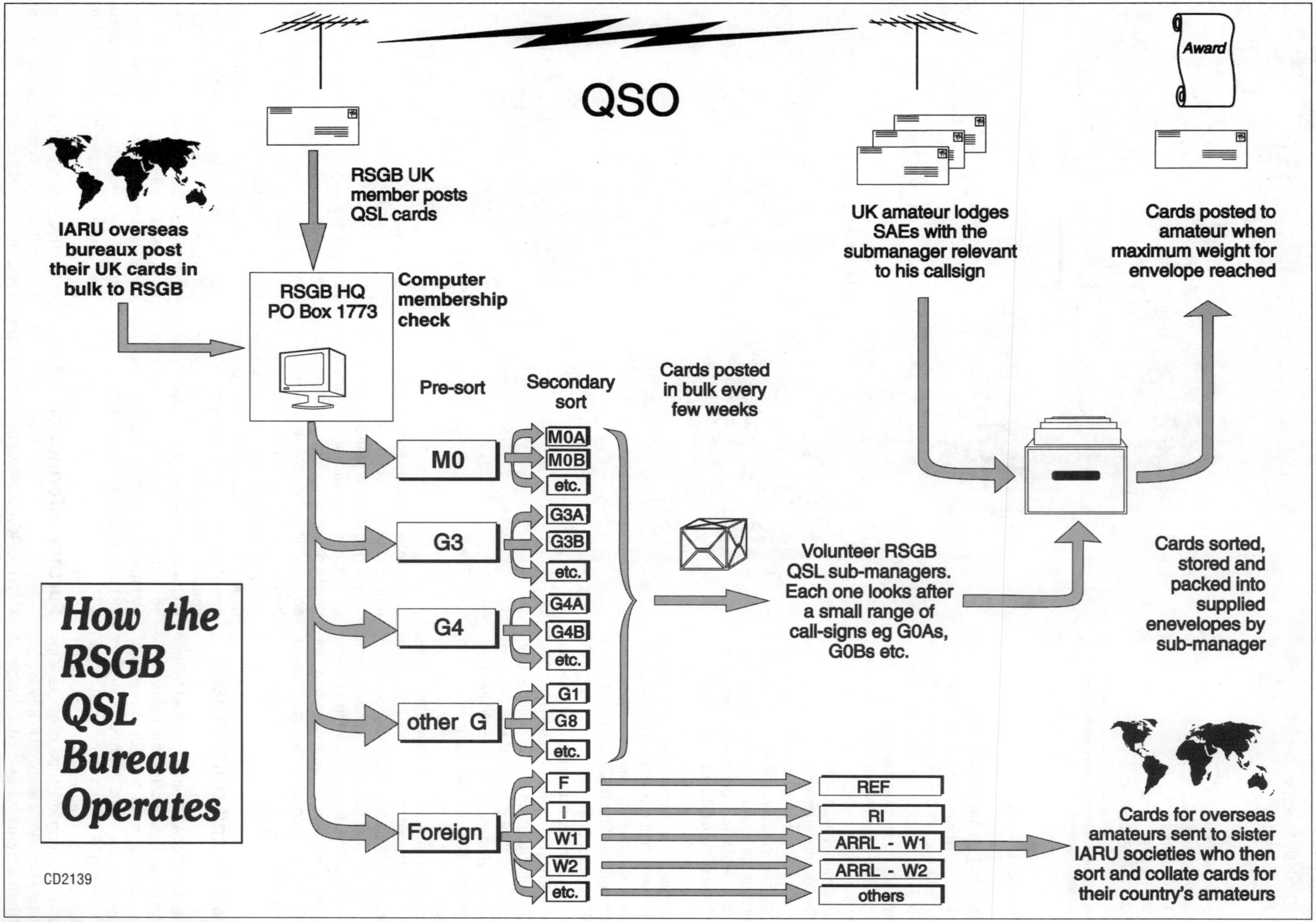

Fig 18: Operation of the RSGB QSL bureau.

Measuring Resistance

IN THIS WORKSHEET you will see how the resistance of a circuit or component can be determined.

There are three methods:

1. Reading the resistance value of a resistor from its colour code markings – you would have seen this if you made the simple DC circuit.

2. Calculation of resistance from known values of current and voltage using Ohm's law – you will do this in a later worksheet.

3. Measurement of resistance directly using a multi-meter as an ohmmeter.

The colour code (10.c1)

RESISTORS ARE normally marked with coloured rings or bands starting at one end of the resistor. This is illustrated in **Fig 19**. The colours of the first three bands represent the numbers relating to the value of the resistance.

There is also a fourth band that indicates how close to the marked value you can expect the actual value to be. This difference is known as the tolerance. A resistor with a 1% tolerance can be expected to be very close to what the coloured bands say, whereas a 10% resistor could be 10% lower or 10% higher than the marked value. For most amateur radio purposes, 5% or 10% tolerance resistors are perfectly adequate.

Some resistors have a fifth band, but you can ignore this at present.

How does the colour code work? The first two coloured bands represent the first two numbers of a resistor's value, while the third band indicates the *number of zeros* that follow these numbers. The first number is read from the coloured band

Black	0 (zero)
Brown	1
Red	2
Orange	3
Yellow	4
Green	5
Blue	6
Violet	7
Grey	8
White	9

Table 4: The resistor colour code. Please note that this should be memorised for the examination

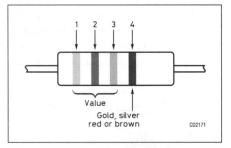

Fig 19: Typical resistor with four coloured bands. Colour examples of actual resistors can be seen on the inside back cover.

that is closest to one end of the resistor. The values for each colour are shown in **Table 4** and in colour on the inside of the back cover of this book.

There are several possible values for resistor tolerance, but the two most common values are worthwhile remembering. A silver band is 10% and a gold band is 5%. This is not difficult to remember if you know that 5% resistors are more expensive and gold is more valuable than silver.

So, if you have a resistor marked 'brown, black, red, silver', its value would be 1 (brown), 0 (black) with 2 (red) zeros. This equates to 1,000 ohms, which is more likely to be written as 1kΩ (remember, 'kilo' means one thousand). The silver band indicates 10% tolerance, so the actual value could be anything between 900Ω and 1.1kΩ (sometimes 1.1kΩ is written as 1k1).

There are a couple of special cases that you should be aware of. Firstly, you may find black as the third coloured band. As black means 'zero' it means that there are no zeros following the first two numbers; a resistor marked: brown, red, black, would be a 12Ω resistor.

For even smaller values of resistor you will find gold used as the *third* coloured band. In this position gold means 'divide by ten'. So, for a resistor marked: yellow, violet, gold, gold, the value will be 47 divided by 10, which is 4.7Ω (sometimes 4.7Ω is written as 4R7), and the actual value will be within a 5% tolerance.

Using the multi-meter as an ohmmeter (10d.1)

YOU ALREADY KNOW that the current flowing through a resistor is related to the voltage across the resistor and the value of the resistor. This means that if we con-

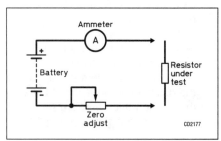

Fig 20: The circuit of a simple ohmmeter.

nect different resistors into a circuit with a battery and an ammeter, the ammeter will indicate different currents. We could therefore replace the current scale on the meter with one marked in Ω.

Fig 20 illustrates how an ohmmeter works. When you select the Ω ranges on a multi-meter, a battery is switched into circuit and the scale reads the current flowing through the resistor. The scale is calibrated so that you can read the measurement in Ω. See **Fig 21**.

This is more obvious with an analogue meter than with a digital meter. If you select a resistance range and touch the test probes together the pointer will move all the way across the scale. As there is no resistance between the probes the internal battery will cause maximum current to flow. You will therefore find zero Ω at the right hand end of the scale. With the test probes apart they are insulated from each other there is a very high resistance between them, so the meter's pointer will not move. Maximum resistance, infinity (often marked '∞'), is therefore at the left hand end of the meter's scale.

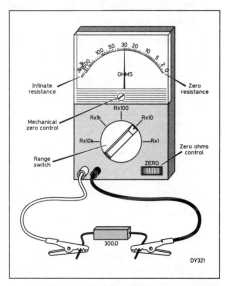

Fig 21: Using a multi-meter to measure resistance.

Practical exercise – resistors in series

1. Take three resistors of different values. Read the colour codes and note down the values.

2. Solder the three resistors between some pins on your DC circuit board or some other suitable surface, so that the resistors are in a straight line. In this configuration they are said to be connected in series (see **Fig 22**).

3. Set the ohmmeter on a suitable resistance range, that is a range that will easily read up to the highest expected value of the resistor (estimate the value from the colour code).

4. Calibrate the ohmmeter by touching the probes together and adjusting the meter to read zero Ω (not usually required with a digital meter). You will find more adjustment is necessary as the internal battery runs down, which it does over a period of time.

5. Place the test probes across each resistor in turn, noting the values. Check these against the colour code values. Double-check any that do not agree, allowing for the tolerance.

6. Add together the three values you have read and measured, and note down the total. Resistances in series are *added* together.

7. Place the test probes across the ends of the three resistors in series and note the total resistance measured. Check this against your calculated total value. Hopefully they will be the same.

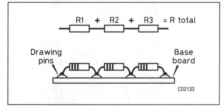

Fig 22: Three resistors in series.

Practical exercise – resistors in parallel

1. Take two resistors of the *same* value. Measure them using the ohmmeter and note down their values. If you are using an analogue multi-meter, don't forget to 'zero' it before use.

2. Solder the resistors together so that they are 'side by side'. These are said to be connected in *parallel* (see **Fig 23**).

3. Measure their total resistance. The total value should be *half* the value of *one* of the resistors, within tolerance.

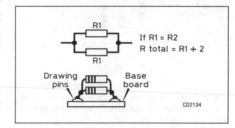

Fig 23: Two resistors in parallel.

Series and parallel (3b.4)

WHY DID THE resistors combine to give a larger value in the first exercise and a lower value in the second exercise?

Resistors joined end-to-end are said to be *connected in series*. The values of the individual resistors are *added* together to give the total value. In a series circuit the current has to flow through each resistor in turn, each one opposing the flow of current and making its passage more difficult. It is therefore no surprise that the resistance values add up.

Resistors joined side-by-side are said to be *connected in parallel*. In a parallel circuit the current has more than one path that it can flow through, reducing the opposition to the flow of current and making its passage easier. Again, it becomes clear that there is *less* resistance than if there had only been one resistor.

If two resistors of the *same* value are connected in parallel, the total resistance will be *half the value of one of them*. If three resistors of the *same* value are connected in parallel, the total resistance will be a *third* of one of them, and so on. Incidentally, this 'rule' only works if the resistors are of the same value. It is possible to calculate the total resistance for resistors of different values connected in parallel, but the maths is beyond the Intermediate syllabus.

As an example, suppose you had a 10V battery and two 100Ω resistors. Connecting one resistor across the battery will allow a current of:

I = V/R = 10/100 (or 0.1A)

to flow. If the second resistor is connected across the first, so that the two resistors are in parallel, 0.1A flows through each resistor and the total current drawn from the battery is doubled to 0.2A. The 10V battery now has 0.2A flowing, so it sees a resistance of:

R = V/I = 10/0.2 = 50Ω

50Ω is half the value of the individual resistors.

Summary

THE MAIN POINTS to remember from this exercise are:

- Resistance is measured in Ohms (Ω, kΩ or MΩ).
- Resistor values can be read by using the resistor colour code.
- Resistance can be measured using an ohmmeter.
- An ohmmeter has an internal battery and must be 'zeroed' before use.
- Resistor values connected in series are added together.
- Two resistors of the same value connected in parallel have a total value of *half* of one of them.

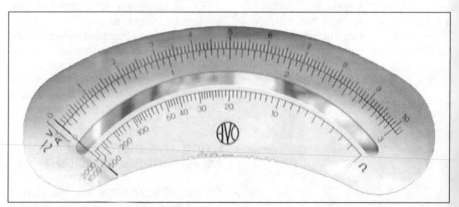

The scale of an AVO8. The resistance scale is the lowest one. Note that zero is on the right hand side.

Capacitors, Inductors & Tuned Circuits

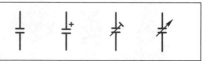

Fig 24: The circuit symbols for fixed, polarised (electrolytic), pre-set and variable capacitors.

Capacitors (3d.1)

CAPACITORS ARE very important components in radios. They can be found in tuning circuits and in a wide range of filters. All capacitors are formed by having two metal plates separated by an insulating material. The insulation is known as the 'dielectric' and circuit symbols are shown in **Fig 24**.

Capacitance (3a.1 & 3d.2)

IF A VOLTAGE source such as a battery is connected to a capacitor, a small current will flow into the capacitor to charge up the plates. This charge will remain, even when the voltage source is removed.

The ability to store an electrical charge is known as 'capacitance'. The unit of capacitance is the Farad (F) but in amateur radio circuits the values used are generally much less than a Farad - micro-Farads (μF), nano-Farads (nF) and pico-Farads (pF) being more common. You will recall that micro (μ) means 'one millionth', but nano (n) and pico (p) are even smaller!

The value of a capacitor is determined by the size of the metal plates and by their separation. Capacitance is normally marked in plain numbers but some use codes like resistors (e.g. 103 means '10 with 3 zeros' = 10,000pf = 10nF)

You need to be able to deal with these sub-units of capacitance and it is worth remembering that:
- 1000pF   = 1nF
- 1000nf   = 1uF
- 1000000uF = 1F

What do capacitors do? (3d.3)

BECAUSE THE METAL plates in a capacitor are separated by an insulating material, a capacitor cannot pass direct current. However, because alternating current changes polarity very quickly, the capacitor will charge on one side and then the other, affectively allowing AC to pass. Capacitors therefore *block DC* and *pass AC*.

As you will see later, the charging and energy storing property of capacitors allows them to be used in tuned circuits and power supplies.

Capacitors in use (3d.4)

CAPACITORS ARE classified according to their insulating material; ceramic, polystyrene, polyester and air being amongst the more common. Capacitors employing these materials tend to have values in the nF and pF ranges. Whilst the shapes and sizes of capacitors vary enormously, all have two connecting leads attached to the metal plates. These types of capacitor (see photo left) can be soldered into circuits either way round.

Some capacitors are 'polarised', these being known as 'electrolytic' capacitors. They tend to have values in the μF range and are shown in the photo bottom left. Be careful when fitting electrolytic capacitors, as they *must* be connected the correct way round (indeed they may explode if fitted incorrectly!). You should see either a '+' or '-' sign against at least one of the leads to help you get it right.

Sometimes we need capacitors with variable values. These generally use air as the insulator and have two sets of metal plates, one set of which can be rotated to vary the amount of overlap, effectively making the capacitor's plates larger or smaller. Variable capacitors are found in 'tuned circuits' and can take the form of pre-set 'trimmers' (see photo immediately below) or externally adjustable 'tuning capacitors' (see photo bottom right).

Typical fixed-value non-polarised capacitors.

Typical pre-set capacitors.

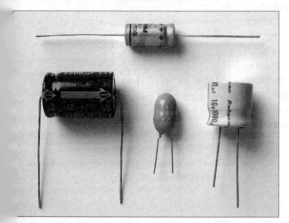

Typical polarised (electrolytic) capacitors.

Typical variable capacitors.

Typical small inductors.

Dangers from stored charges (9d.4)

In circuits with high potential differences present capacitors may store dangerously high electric charges for long periods of time. Good circuit design should ensure that these charges leak away safely when the equipment is switched off. However, if you are working on 'high voltage' equipment any capacitors likely to store dangerous charges must be discharged safely before starting work.

What is an inductor? (3e.1)

AN INDUCTOR IS normally formed from a coil of wire, with the turns insulated from each other. The circuit symbols for various types of inductor are shown in **Fig 25**. Like capacitors, inductors are used in tuned circuits and filters. Without capacitors and inductors it would be difficult to have any radio communication at all!

Electro-magnetism and inductance (3a.1, 3e.2 & 3e.3)

WHEN A CURRENT flows through a wire, a magnetic field is induced around it. If a wire that is carrying current is wound into a coil, the magnetic field is enhanced. An inductor is able to store energy in its magnetic field, this property being known as *inductance*. This property is used to form electro-magnets.

The unit of inductance is the Henry (H). In radio we tend to use much smaller values than the Henry, the milli-Henry (mH) or the micro-Henry (μH) being most common.

The value of an inductor is determined by its diameter, the length and number of turns in the coil. The material upon which it is wound (the core) also affects the value.

A large variable inductor.

Inductors in use

FIXED VALUE INDUCTORS come in a wide range of shapes and sizes. One of the more popular types found in radios uses a 'toroidal' core, which looks like a small ring. These are quite easy to wind and are commonly used in kits. The right hand component on the middle row of the photo top left shows a toroidal inductor.

Variable inductors come in two main types. The 'roller coaster' (pictured above) has a sliding contact that moves along the length of the coil as it is rotated, selecting more or less of the turns (the same effect can be accomplished by switching between 'taps' on a fixed inductor). These are commonly used in antenna tuning units. A more common type of variable inductor has a core made from a material known as ferrite, which can be moved into the coil on a screw thread. These are often found in tuning circuits and an example is shown in the photo top left (left hand component on the middle row).

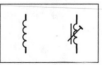

Fig 25: The circuit symbols for fixed and variable inductors.

FARAD

Unit of capacitance is the Farad. Values of a Farad and milli-Farad are not normally used in radio circuits.

1F = 1000mF (not normally used)
1mF = 1000uF (normally expressed in uF)
1uF = 1000nF
1nf = 1000pf

Note - to convert from one level to the next, move the decimal point 3 places

HENRY

Unit of inductance is the Henry. Values of a Henry are not normally used in radio circuits.

1H = 1000mH
1mH = 1000uH

Note - to convert from one level to the next, move the decimal point 3 places

Capacitors, Insuctors & Tuned Circuits (Worksheet 15)

Tuned circuits (3f.1)

WHEN A CAPACITOR and an inductor are joined together they form a 'tuned circuit'.

If we assume the capacitor is charged, its energy will flow into the inductor, which will store the energy in its magnetic field, discharging the capacitor in the process. The energy in the coil then transfers back to the capacitor, charging it up again.

This is a bit like an electrical version of a child's swing. At the top of its travel energy is stored as potential energy. This encourages it to descend, gathering speed as it does so. By the time the swing reaches the bottom of its travel the potential energy has all been converted to kinetic energy – the energy of motion. Kinetic energy makes the swing climb up the other side, but as it does so the kinetic energy is converted back to potential energy. This process repeats, but gradually decays until eventually the swing comes to rest.

Resonant frequency (3f.2)

THE FREQUENCY AT which a capacitor and a coil transfer energy back and forth is known as the 'resonant frequency'. At the resonant frequency, energy will be transferred between the coil and the capacitor very efficiently.

Depending on the quality of the components involved, the circuit may only be resonant on a very specific frequency or it may be resonant over a narrow band of frequencies. At all other non-resonant frequencies the components will act independently.

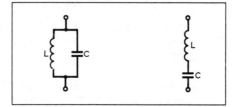

Fig 26: Parallel and series tuned circuits.

Accepting and rejecting current flow (3f.3)

A CAPACITOR AND inductor can be arranged in series or in parallel, as **Fig 26** shows. Both combinations form tuned circuits.

The series tuned circuit is sometimes referred to as an 'acceptor', because it allows current to flow at the resonant frequency, whilst resisting the flow at other frequencies. Such a circuit might be useful at the input to a receiver, to aid selection of the wanted frequency.

The parallel tuned circuit is sometimes referred to as a 'rejector', because it resists current at the resonant frequency, whilst allowing the flow at other frequencies. One place that such a circuit is found is in a trap dipole (**Fig 27**), to isolate the outer sections of the antenna at the resonant frequency of the traps.

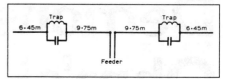

Fig 27: A trap dipole antenna contains two parallel tuned circuits.

Filters (4e.4)

CAPACITORS AND inductors are also used in various combinations to form filters, i.e. circuits that block some frequencies and pass others. You will meet the following types later in this book:

- Low Pass
- High Pass
- Band Pass

Demonstrating Ohm's Law

IN THE Foundation training you met Ohm's Law. That rule states that 'the current through a resistance is proportional to the voltage across it'. This means that an increase in potential difference across a resistance will cause an increase in the current through the resistance.

As you will recall, the formula derived from Ohm's Law is:

$$V = I \times R$$

where V is the potential difference in volts, I is the current in amps, and R is the resistance in ohms (Ω). The equation can be re-arranged so that you can calculate the potential difference, the current or the resistance, so long as you know the other two values.

The triangle shown in **Fig 28** is an easy way of remembering the appropriate equation. Place your finger over the symbol you need to calculate and the two remaining ones tell you how to use them; side-by-side means multiply, one over the other means divide. If you are in any doubt about which letter goes where, remember this is a *Very Important Rule*. The 'V' from *Very* sits at the top of the triangle.

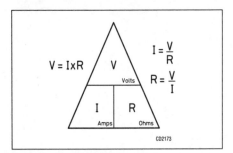

Fig 28: Cover the letter of what you want to know and the remaining letters show you how to arrange the formula.

Using the DC circuit (3b.4)

DOES THE THEORY work in practice? In the following exercise you will measure the current flowing through the 470Ω resistor (R2) and the potential difference across it. Using Ohm's Law you will calculate the actual value of the resistor and then you can check the calculation by measuring the resistance using the ohm-meter, as shown in **Fig 29**.

1. Prepare the DC circuit by removing the shorting wire from the 470Ω resistor (R2).

2. Set the multi-meter to a suitable direct current range (mA, DC) and with the test probes correctly orientated, measure the current through the circuit by placing the probes across points B and C. Record the reading. Refer back to Worksheet 10 if you are unsure how to do this.

3. Remove the multi-meter and turn the circuit on by closing the switch at B-C.

4. Set the multi-meter to a suitable DC volts range and measure the potential difference across the resistor R2, then turn the circuit off again. Refer back to Worksheet 9 if you are unsure how to do this.

5. Using the Ohm's Law triangle, calculate the resistance of R2 from your measurements. Covering R leaves V over I, so you need to divide the potential difference by the current. If you measured the current in milliamps, remember to convert it to amps first. Do this by moving the decimal point three places to the left. Record your result.

6. Make sure the circuit is 'off', set the multi-meter to a suitable resistance range, place the probes across the resistor R2 and record the value. Refer back to Worksheet 14 if you are unsure how to do this.

How does the ohmmeter reading compare to the calculated value? They should be very close.

You will remember that when you built the DC circuit, R2 was selected using its coloured bands (yellow-violet-brown), 470Ω. How do the measurements compare to the marked value of the resistor? There may be some difference this time, because the resistor tolerance allows a fairly wide variation. With a 10% tolerance, the resistor could be anything between 423 and 517Ω. Is it?

As a crosscheck, using your measured value of resistance and potential difference, use the triangle to calculate the current. Remember, this is a small current and the units are small (milliamps). Compare the calculated value to the measured value. Are they close?

Summary

THE MAIN POINTS to remember from this exercise are:

- You have calculated the resistance from measured values of voltage and current
- You have measured the same resistance using an ohm-meter and found it to be fairly close to the calculated value.
- You have calculated the current from measured values of resistance and potential difference.
- The figures were also close to those expected, showing that the theory of Ohm's Law works in practice… if you know two of the values, you can calculate the third.

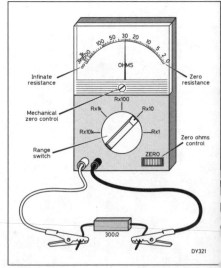

Fig 29: Use of an ohm-meter to measure resistance

RF Oscillators

Types of RF oscillators (4b.1)

AS YOU MAY recall from your Foundation transmitter training, the frequency generator, or RF oscillator, produces the radio frequency energy that will be transmitted. As you will find out later in this training, some receivers include oscillators too.

All oscillators need to be very stable, that is to say they will not drift or be pulled off frequency unless we want them to.

There are two main types of RF oscillator; crystal oscillators and variable frequency oscillators (VFOs). You need to know the key features of each, together with their advantages and disadvantages.

Crystals (3i.1)

QUARTZ CRYSTALS can be found in oscillators and filters. The mineral has a particular electro-mechanical property known as the 'piezoelectric effect'. Each crystal has a specific frequency at which the effect is most pronounced. The frequency of the crystal is determined by various factors such as its size and thickness. The circuit symbol for a crystal is shown in **Fig 30**.

A crystal is usually housed in a metal can, with two connecting leads or pins. Some crystals are designed to be soldered into circuits, whilst others have pins which fit into sockets, enabling quick changes if required.

Crystal oscillators (3h.11)

IF A CRYSTAL is used together with an active device, such as a transistor, it will form a very stable oscillator, with the crystal setting the operating frequency.

Crystal oscillators tend to be very stable and produce good quality signals. However, the main disadvantage of a crystal oscillator is that it only operates on one frequency. If you want to change frequency you will need another crystal. This arrangement used to be very popular for the RF generator in VHF and UHF transmitters, where contacts take place on specific channellised frequencies. The circuit diagram of a typical crystal oscillator is shown in **Fig 31**.

Fig 30: The circuit symbol of a quartz crystal.

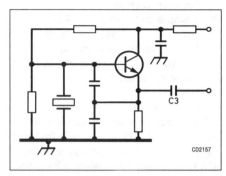

Fig 31: The circuit diagram of a Colpitts crystal oscillator.

Variable frequency oscillators (3h.11, 4b.2 & 4b.3)

A MORE USEFUL oscillator for wide ranges of frequencies is the Variable Frequency Oscillator or VFO (see **Fig 32**). Don't worry if some of the circuit symbols are unfamiliar, all will be revealed very soon! However, if you compare Fig 31 to Fig 32 you should notice that the crystal is replaced by a tuned circuit formed by an inductor and a number of capacitors. The variable capacitor used in the tuned circuit allows the frequency to be varied over a range of frequencies. You do not need to remember the full circuits but you may be asked to recognise crystal and inductor/capacitor based oscillators.

The range of a VFO will be determined by the minimum and maximum values of the variable capacitor. For example, a 10–100pF capacitor will enable a wider range to be used than a 10-25pF component. When setting the frequency of a VFO, you may also need to vary the inductance by adjusting a threaded core at the centre of the coil. It sounds trickier than it is and you will learn more in the VFO exercise later.

The VFO may appear to offer a much better service than the crystal oscillator, but it has a number of disadvantages. The first is that it may drift or be pulled off the frequency that you set it to. A VFO requires careful construction to prevent it from drifting. Features of a good VFO include:

- A well-regulated DC supply. Changes in supply voltage can cause changes in operating frequency
- Rigid mechanical construction. Any mechanical movement of the components can cause a change in the tuned circuit and hence the operating frequency

A selection of quartz crystals.

- A screened enclosure. Placing a metal screen around the VFO circuit will prevent it from being affected by rapid temperature changes and stray RF from other circuits
- Using a buffer amplifier. A buffer amplifier effectively isolates the VFO from the rest of the radio circuitry, so any changes – e.g. switching from receive to transmit – will not affect the VFO

The second problem which arises from the use of a VFO is the potential for transmissions outside of the amateur bands. If a VFO is setting the operating frequency of your transmitter and it is capable of oscillating beyond the edges of the amateur bands, you could inadvertently transmit on the frequency of another radio user service; a clear breach of your licence conditions.

Therefore it is crucial that a VFO is calibrated to show the amateur band edges as a minimum. In practice you will probably want to calibrate it so that you know exactly what frequency it is set to.

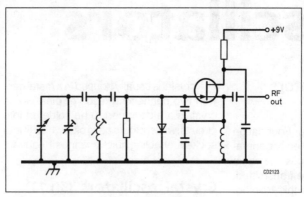

Fig 32: The circuit diagram of a Colpitts VFO.

Digital frequency synthesisers (4b.4)

MOST MODERN transmitters use oscillators known as digital synthesisers. These are based on crystal oscillators, but enable a wide range of frequencies to be selected. They are very stable but can be quite complex to build. For that reason, most 'first time' kits and designs use standard crystal oscillators or VFOs.

Calibrating a VFO (10f.1)

THERE IS AN optional practical exercise included in this book. If you have never calibrated a VFO it is certainly worth doing. You will be required to demonstrate your ability to calibrate a VFO as part of your practical assessments, so even if you don't complete this practical exercise now, you should be familiar with the procedure.

Calibrating a VFO

Practical assessment (10f.1)

PART OF THE practical assessment requires you to demonstrate the procedure for calibrating a VFO. You are not required to *build* a VFO, but you will have to satisfy your assessor that you can calibrate one. A VFO circuit was shown in the RF Oscillators worksheet of this book and a physical layout for that oscillator is provided in the exercise below. There is no compulsion for students to build a VFO but the layout is there in case you would like to build an oscillator as well as calibrating one.

If you are taking part in a tutor-led course, completing the optional exercise could count towards the practical assessment. Alternatively, if your Intermediate training project includes a VFO, you may use that and follow the appropriate instructions.

Test equipment

RF ENGINEERS have a wide range of sophisticated test equipment to help with calibration tasks. However, most amateurs have to make do with a less complex test bench.

It is possible to use a digital frequency meter that displays frequency directly, but these are not widely available and may not be as accurate as the precise readout leads you to believe.

The most readily accessible calibration tool is a receiver which is capable of covering the range of the VFO under test. As well as being a useful means of listening to the amateur bands, a general coverage receiver is an excellent item of test equipment.

Even the best screened VFO will radiate a low level signal and as receivers are designed to pick up very weak signals it should be possible to listen to the VFO and measure its frequency on the receiver's display or dial.

First of all, select the approximate frequency on the receiver and set the mode to CW or SSB. You need to tune to the VFO signal until the pitch reduces and stop when you cannot hear the VFO any more. This is known as 'zero beat', the point where there is no difference between the VFO and the receiver's frequency. But how do we know if the receiver is showing the right frequency?

The receiver itself can be calibrated against a signal of known accuracy, such as the standard frequency services (e.g. WWV from the USA on 2500, 5000, 10000, 15000 and 20000kHz), or - as **Fig 33** shows - by a simple item of test equipment known as a 'crystal calibrator'. Some receivers have a crystal calibrator built into them. If you use one of these methods and find that your receiver is constantly 1kHz out, you will know to make suitable corrections when calibrating your VFO.

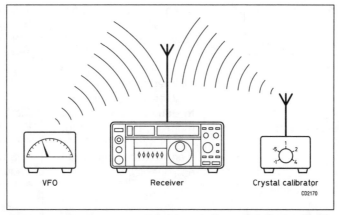

Fig 33: A crystal calibrator and a VFO can be monitored on a general coverage receiver.

VFO adjustments

THE EXACT calibration adjustments will depend on the VFO you are working on, but there are normally two types of adjustments that can be made.

Firstly, the inductor in the VFO's tuned circuit may have a core on a screw thread that can be moved in and out of the coil. This is normally used to set the VFO's mid-range. By setting the variable capacitor to its mid position, the inductor's core can be adjusted whilst listening to the middle of the intended frequency range. Special, non-metallic adjustment tools are made for use with inductor cores. Using an ordinary screwdriver will not only affect the oscillator's frequency but it *could* shatter the core and render it useless. If you do not have a suitable tool you would be well advised to buy one from your inductor supplier.

The second adjustment is normally a pre-set variable capacitor or trimmer that limits the range of the main variable capacitor. Its adjustment may require some trial and error, swinging the VFO through its range, checking the frequencies at each end then adjusting the trimmer to increase or decrease the swing.

Ideally the VFO should only operate within the amateur band edges but it is quite acceptable for it to cover a wider frequency range, providing the dial is clearly calibrated to show the limits of the amateur frequency allocation.

Practical exercise (10f.1)

1. Build the VFO as shown in **Fig 34** and **Fig 35**, or borrow a ready built one from your instructor. This construction is known as 'ugly style', by-the-way. It may not look pretty, but it is quite effective for RF circuits.

2. Set the trimmer so that the moving vanes are fully unmeshed (i.e. not overlapping the fixed vanes) and set the inductor core so that it is level with the top of the screening can.

3. Connect a 9V battery, observing the correct polarity, and place the VFO close to your receiver. You may need to use a short length of hook-up wire as an antenna for the receiver.

4. Tune the receiver to 3650kHz and set the mode switch to receive CW, SSB, LSB or USB. Slowly move the VFO's variable capacitor through its range. Hopefully you should find a clear signal from the VFO. If not, check that all connections are made and try again.

5. Now check the VFO's range by slowly tuning the VFO and following the signal with the receiver. When you have reached the limit in one direction, note the zero beat frequency and repeat this step in the opposite direction. You should now have a record of the VFO's range (e.g. 3440 to 3870kHz).

6. Reset the trimmer so that the moving vanes are fully meshed (i.e. overlapping the fixed vanes) and repeat the previous step. You should find that the VFO's range has changed and you may have to listen either side of the previous range to find the VFO signal.

7. Now unscrew the inductor core by two complete turns and repeat step 5. Again you should find that the VFO covers a different range.

8. Try adjusting the trimmer and the inductor until the VFO covers 3500 to 3800kHz with a minimum of extra coverage (e.g. 3475 to 3825kHz) then mark the band edges on your dial.

Observations

YOU MAY HAVE noticed that the signal from the VFO drifts in frequency slightly when you are listening to it on your receiver, particularly when it is first switched on. This highlights the need to apply good construction practice.

If your VFO does not drift and you need to be convinced about stability, try blowing hot air over the VFO, or shaking it whilst listening to it on your receiver. Gently tapping the VFO with the rubber at the top of a pencil is another test of stability.

This circuit has been tried and tested on many occasions and should work as described, but if you experience difficulties discuss it with your instructor. If you are learning alone, see if you can find a more experienced amateur to help you.

> **VFO component list:**
> **Resistors:**
> 100R = 100Ω = 1
> 100k = 100Ω = 1
> **Capacitors:**
> 1nF disc ceramic = 2
> 10nF disc ceramic = 1
> 470pF polystyrene = 1
> 180pF disc ceramic = 1
> 60pF trimmer = 1
> 150+75pF polycon variable = 1
> **Inductors:**
> TOKO KANK 3334 = 1
> 1mH RF choke = 1
> **Semi-conductors:**
> 1N4148 diode = 1
> J304 FET = 1**
> **Miscellaneous:**
> PCB single sided = 9x12cm
> PP3 battery snap
> PP3 battery
> Knob for variable capacitor
> Rubber feet = 4
>
> ** Any similar FET will work (e.g. 2N3819) but check Source, Gate, and Drain pin configuration with your supplier. Not all FETs have the same pin configuration

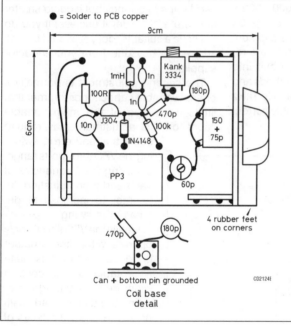

Fig 34: VFO layout.

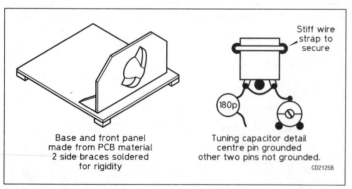

Fig 35: Front panel and tuning capacitor detail.

Semiconductors

BACK IN WORKSHEET 3 we discussed that the electrons in the outer orbits or 'shells' of a conductor move from atom to atom when a potential difference (voltage) is applied across its ends, causing a current to flow. The opposite is true of an insulating material. It has no freely mobile electrons, so no current will flow.

Somewhere between the two there are substances that are neither conductors nor insulators. These are known as 'semi-conductors'. There are a number of these substances, but the one that is most commonly used is silicon – you may have heard of 'Silicon Valley', where many semiconductor companies are located.

Pure silicon is prepared for use in electronic devices by adding other substances to form two different types of impure silicon material. These are known as 'P' and 'N' type materials. The N-type material is said to have 'spare' electrons, whereas P-type material appears to have a few electrons 'missing'.

On their own the two materials are of little use, but when the two are joined together they can form very useful semi-conductor devices for use in radio and electronic circuits.

Diodes (3h.1)

AS ALREADY mentioned in the 'Components' worksheet, a diode allows current to flow in one direction only. This may not seem much, but diodes have many uses.

Perhaps the most common application in amateur radio is for 'rectification' – the process of changing AC into DC. As you will see in later worksheets, this property can be used in a mains power supply or in the demodulator of an AM receiver. Some diodes are made from materials that emit light as they conduct. These are known as light emitting diodes (LED) and are used as indicators.

This all sounds very interesting, but how do the P- and N-type materials allow diodes to work?

If a piece of P-type material is joined to a piece of N-type and two connection wires (electrodes) attached, the completed device is known as a diode. Because of the different types of material used, current will only flow through the device in one direction. When connected to a voltage source, the 'spare' electrons from the N-type material will readily move across the junction to fill the gaps in the P-type material, current will flow and the diode will conduct.

If the voltage source is connected in the opposite direction, the P-type material has no spare electrons to move across the junction, so no current will flow and the diode will not conduct. This 'one-way' property is clear from the circuit symbol (**Fig 36**) for a diode, an arrow pointing in the direction of current flow (conventionally seen as positive to negative, even though it is opposite to the direction that the electrons move).

Even though the current will flow only when the voltage is connected with the correct polarity (the right way round), there has to be a certain amount of electrical potential before any current will flow. We say that the diode requires a small voltage (about 0.6 volts for a silicon diode) to turn it on. Once it is 'on' the diode conducts and has a very low resistance in the 'forward' direction.

You do not have to remember the turn-on voltage, as it varies from diode to diode. Diodes used in crystal diode receivers require much less turn-on voltage, or they would only be able to receive very strong signals. On the other hand, you should know that if the voltage is reversed no electrons will flow at all – the diode becomes an insulator with a high resistance in the 'reverse' direction.

Fig 36: The circuit symbol for a diode.

Transistors (3h.7)

IN AN AMATEUR radio transceiver you will find transistors being used as switches, amplifiers and oscillators. At this level you need to understand a little of how they work.

A transistor can be seen as a sandwich of three pieces of silicon, the two outer pieces being of one type with the middle layer the other type.

If the two outer materials are of N-type silicon, the middle layer will be P-type silicon. This would be called an NPN transistor. If the materials in the sandwich are reversed, then it becomes a PNP transistor. Don't worry too much about the different types at this stage, as only the NPN is used in the Intermediate syllabus. However, you must make sure you use the correct type of transistor if you are building a project.

Wires connected to the layers of silicon in a transistor form the electrodes. The two outer electrodes are called the 'emitter' and 'collector'. The middle electrode is called the 'base'.

In use, a positive voltage is applied to the collector and a negative voltage to the emitter. Electrons will *not* flow until a small positive voltage is also applied to the base. This is approximately 0.6 volts. This small 'turn on' voltage causes a small current to flow from the base to the emitter, which in

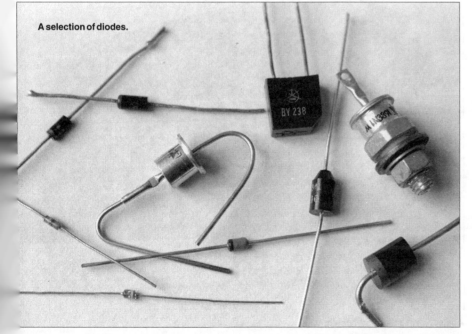

A selection of diodes.

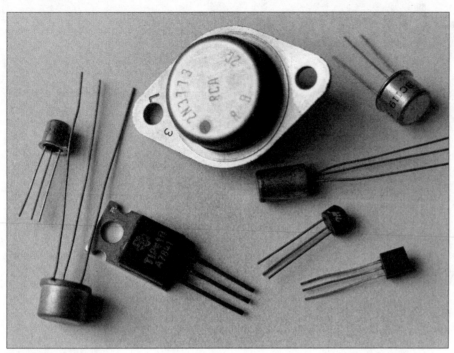

A selection of transistors.

turn causes a much larger current to flow from the collector to the emitter.

This means that the base to emitter current can be used to control the collector to emitter current. This effect can be used to make a transistor operate as a switch, perhaps to change a transceiver from receive to transmit, or to key a CW transmitter.

Now let's say that there is just enough voltage at the base for a small current to flow through to the emitter. If we then apply an additional small AC current to the base, the additional current through the base to the emitter will produce a larger current from the collector to the emitter. Effectively we produce a larger copy of the original AC current. This effect is called 'amplification' and is used for AF and RF amplifiers.

Fig 37: The circuit symbol for an NPN transistor.

Types of transistor (3i.1)

THE TRANSISTOR described previously is known as 'bipolar' (see **Fig 37** for the circuit symbol). A different type is the 'Field Effect Transistor' or 'FET' (see **Fig 38** for the circuit symbol). These are also made from P- and N-type materials, but their construction gives them different properties. The electrodes of the FET are known as the 'source', 'gate' and 'drain'.

Don't worry too much about the different types of transistor at this stage, just be able to recognise the symbols and know the names of their connections. Be sure to fit the correct types the correct way round in your projects.

Fig 38: The circuit symbol for a FET.

Integrated circuits

MORE AND MORE circuits now use integrated circuits (ICs). These are also electronic devices based on P- and N-type materials, but they can contain hundreds or thousands of transistors and diodes.

Some ICs are made for general use, perhaps forming a block of four or five transistors in one device, whereas others have very specific uses; a mixer, an amplifier (see **Fig 39**) or even a whole radio receiver.

Many ICs come in 'dual in-line' packages with many connection pins. A dot on the top of the package usually indicates which is pin number one, providing a useful guide for you to ensure you solder it into the circuit the right way round.

Fitting an IC incorrectly can damage it, so take care!.

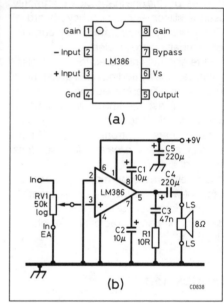

Fig 39: An LM386 audio amplifier, showing that only a few external components are required.

A selection of Integrated Circuits (ICs).

Using Diodes

IN THIS WORKSHEET you will use two different diodes with the simple DC circuit, to learn more about their properties.

Getting to know diodes (10d.5)

FOR THIS EXERCISE you will need a multi-meter, your simple DC circuit and two diodes; a silicon diode (e.g. a 1N4148 or a 1N4001) and a small red light emitting diode (LED). See **Fig 40** for circuit symbol.

1. Set your multi-meter to its highest resistance range. Note that this may not work with some digital multi-meters. If you have one, select the 'diode test' position. If not, move on to **'Find out more'** below.

2. Measure the resistance of the silicon diode. You will note that the diode has a band around one end. Measure it first with the band towards one test probe and then towards the other. Record the results (if both results are the same, something is not quite right, so go back and check).

3. Repeat the resistance measurement with the LED. This time there is no band to indicate which end is which, but you should find that one connecting wire is longer than the other and/or there is a flat side on the body of the LED. Again, record the results.

You should have found that the diodes both had high resistances when tested in one direction (this we call the 'reverse') but much lower resistances in the other direction (this we call the 'forward'). In other words, in one direction they are conductors, in the other they are insulators.

Find out more

1. Replace the shorting wire across the 470Ω resistor (R2) that was removed in Worksheet 16.

2. Take the DC circuit and check that all is well. Switch it on by shorting B to C. The bulb should light. Now switch it off again.

3. Take the diode and touch it across points B and C, first with the band towards B, then towards C. You should find that the bulb only lights when the diode is connected one way. Again it seems that the diode is both a conductor and an insulator, depending on which way it is fitted. Make a note of which way this is.

4. Repeat the exercise with the LED. Record which way round it lights.

Fig 40: The circuit symbol for a LED.

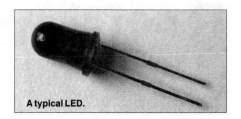

A typical LED.

You should have found that the bulb did not light with the LED either way round but that the LED did light in one of the tests. Why did the LED light but the bulb not? Current must have been flowing for the LED to light, so why did the bulb not light? No answers just yet, have a think.

Modifying the DC circuit

LET'S MAKE the LED do some more work by fitting it into the DC circuit (**Fig 41**).

1. Remove the bulb and the shorting wire across R2.

2. Close the switch between B and C.

3. Touch the LED electrodes across the bulb holder connections. The LED will light with its electrodes only one way round. Make a note of their position.

4. Turn off the switch B-C.

5. Fix the LED into position across the bulb holder's terminals.

6. Check that the LED lights when the switch B-C is closed.

7. Measure the voltage across the LED.

8. Open the switch B-C and measure the circuit current at this point.

9. Calculate the resistance of the LED from the formula R = V/I (check the Ohm's Law triangle if you have forgotten). Note the result.

You will notice that the calculated forward resistance of the LED is different to your measured value. Why should that be?

Some answers

THE FIRST PUZZLE in these exercises was why the LED lit with the bulb in circuit when the bulb did not. The answer lies in the current required to make an LED light. Compare your results here with the current you measured in the earlier exercise when the bulb was lit. It should be clear from this that LEDs require much less current to light than bulbs do. The extra series resistance of the LED reduced the current flowing in the circuit, so current was flowing through the bulb, just not enough for it to light.

The second question relates to the different forward resistance values for the LED. When you did the exercise with a resistor, the result of the Ohm's Law calculation was quite close to the measured value, but not with the LED.

The reason for this is that the voltage across the LED in the circuit is different to that voltage across the ohmmeter probes (remember the ohmmeter has an internal battery). Diodes and other semiconductors do not have fixed resistances, they change depending on the current flowing through them.

Summary

- An LED will light when a current flows through it in one direction only.
- Like an LED, a silicon diode conducts in one direction only.
- Both diodes have a relatively low resistance in the forward direction compared to the reverse direction.
- The forward resistance of the LED was different when in the circuit, compared with the measured ohmmeter value.
- Semiconductors do not have fixed resistances.

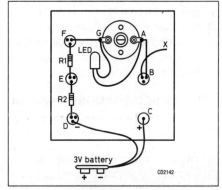

Fig 41: Layout of the modified DC circuit.

Revision Questions 1

YOU HAVE NOW reached the halfway stage in this training course and it is time to reflect on what you have learned so far. One of the best ways to do this is to see if you can answer a few questions like those you will meet in the written exam.

These are not 'official' questions from the exam question bank, but they are similar in style. They are intended to set the scene for the real exam at the same time as encouraging you to look back over the first half of the book. Good luck!

1. Which ingredient in cored solder helps it to flow freely on a joint?
   (a) Lead
   (b) Tin
   (c) Flux
   (d) Copper

2. Which of the following components does NOT need to be fitted the right way round?
   (a) A diode
   (b) A resistor
   (c) A transistor
   (d) A logarithmic potentiometer

3. The flow of electrical current can be described as the movement of:
   (a) Electrons
   (b) Neutrons
   (c) Protons
   (d) Atoms

4. The Earth wire on a 3-core mains lead is:
   (a) Brown
   (b) Blue
   (c) Black
   (d) Green and yellow

5. When measuring direct current, a multi-meter should be:
   (a) Connected in parallel to the component under test
   (b) Switched to a high voltage range
   (c) Switched to a low resistance range
   (d) Connected in series with the circuit under test

6. What would be the MOST likely effect of regulating the DC supply to a VFO?
   (a) The oscillator would generate harmonics
   (b) The oscillator would change frequency
   (c) The oscillator would be modulated
   (d) The oscillator would be more stable

7. Which of the following means 'interference from another station'?
   (a) QSO
   (b) QRT
   (c) QRM
   (d) QRN

8. The fuse in a 3-pin mains plug is:
   (a) Designed to fail when current passes through it
   (b) Designed to protect the user from electric shocks
   (c) Made from a good insulating material
   (d) A metal strip that fails under fault conditions

9. The value of a capacitor is measured in:
   (a) Henry
   (b) Hertz
   (c) Farads
   (d) Siemens

10. Which of the following oscillators would be LEAST stable?
    (a) Variable Frequency Oscillator
    (b) Crystal Oscillator
    (c) Digital Frequency Synthesiser
    (d) Variable Crystal Oscillator

11. Which of the following stations is located in Scotland?
    (a) M3QQQ
    (b) MW3QQQ
    (c) 2E0QQQ/M
    (d) 2M0QQQ

12. Which of the following could be used to indicate that a radio was switched on?
    (a) A zener diode
    (b) A silicon diode
    (c) A varicap diode
    (d) A light emitting diode

13. The frequency of a tuned circuit will NOT be affected by:
    (a) The area of the capacitor's plates
    (b) The number of turns in the inductor
    (c) The core of the inductor
    (d) The type of capacitor (e.g. ceramic or air-spaced)

14. When referring to soldering, the term 'dry joint' means:
    (a) A form of soldering using dry paste
    (b) A good solder joint
    (c) A joint that does not conduct current
    (d) A poor joint that is likely to fail

15. A bipolar transistor has three connections. These are:
    (a) Anode, cathode and grid
    (b) Anode, collector and base
    (c) Collector, base and emitter
    (d) Cathode, gate and junction

16. Which of the following is an insulator?
    (a) Copper
    (b) Aluminium
    (c) Carbon
    (d) Polystyrene

17. If two resistors, each one being 50kΩ, were connected in series, what would the total resistance be?
    (a) 1kΩ
    (b) 25kΩ
    (c) 50kΩ
    (d) 100kΩ

18. How many kHz is 3.5MHz?
    (a) 35
    (b) 350
    (c) 3,500
    (d) 35,000

19. An Intermediate Licence holder is allowed to operate:
    (a) From a boat on an inland canal
    (b) In any other country that has an amateur service
    (c) From an aircraft
    (d) From a ship in the Atlantic ocean

20. The frequency of a crystal oscillator is mainly determined by:
    (a) The physical properties of the crystal
    (b) The transistor being used in the circuit
    (c) The size of the crystal holder
    (d) The DC supply voltage

How did you get on?

IF YOU FIND THAT you cannot answer some of these questions, go back and read through the appropriate worksheet(s) again. If you are still stuck, see your instructor or ask a more experienced amateur to help.

It is more important to be confident about the topics than to get the right answers. Remember, these are not exam questions, but the topics covered are from the syllabus. Once you are happy with the material covered so far, move on to the second half of the book.

Transmitters

Block diagrams (4a.1 & 4a.2)

THE FOUNDATION syllabus includes a simple model of a transmitter using four blocks, the audio stage, the frequency generator (oscillator), the modulator and the power amplifier. We are now going to add a little detail to the Foundation model, to see how the following types of transmitters work:

- Morse (CW)
- Amplitude Modulated (AM)
- Frequency Modulated (FM)
- Single Sideband (SSB)

Before we look at the different transmitters you should understand what each of the blocks is there to do. Some of the blocks are the same as those used in the Foundation model, but there are also some new ones to learn about.

There are four blocks that are the same in all the transmitters that use them, so we will look at those first. The blocks which are specific to a particular transmitter will be dealt with as they occur.

RF oscillators (4b.1)

YOU SHOULD ALREADY have covered the different types of oscillators in an earlier worksheet. Any of the oscillators discussed could be used in a transmitter. Do you recall their advantages and disadvantages?

Power amplifiers

ANOTHER BLOCK you have met before. At this level you don't need to know any more about it than you learned at Foundation level - it produces a larger copy of the modulated signal.

Low pass filters (4e.3, 4e.4 & 4e.5)

EVERY TRANSMITTER should include a low pass filter (LPF) as the last block before the antenna connection. The LPF prevents the radiation of harmonics, which are multiples of the frequency that you intend to transmit. Harmonics can be generated in any active part of the transmitter, particularly in amplifiers. They can cause

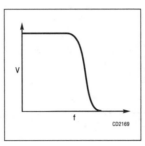

Fig 42: A low pass filter permits frequencies up to a certain value to pass through it, but higher frequencies are blocked.

interference to other amateur bands and other radio users. We will come back to look at harmonics later (Worksheet 34).

There is no such thing as the 'first harmonic', the design frequency is normally referred to as 'the fundamental', the second harmonic is twice the fundamental and the third harmonic three times, etc. The harmonics from a 51MHz transmitter would therefore appear at 102, 153 and 204MHz. None of these are amateur frequencies, so you would be in breach of your licence conditions if you radiated on these frequencies. The third harmonic is generally seen to present the highest risk of interference.

As **Fig 42** shows, the LPF is designed to stop frequencies above a cut off point, but pass frequencies below that point. For example, the LPF in a 3.5MHz transmitter might be designed to cut off anything above 4.0MHz, so the 3.5MHz signals would pass through but any 7, 10.5 or 14MHz signals would not.

Microphones

ALTHOUGH a microphone (see **Fig 43** for circuit symbol) was shown as part of the Foundation transmitter model it is included here to explain a little about how it works.

A microphone converts mechanical air movements, a person speaking for example, to electrical signals at audio frequency (AF). This is done by a small diaphragm that is connected to a coil of wire in a magnetic field. The movement of the coil in the field causes current to flow in response to the mechanical movements.

Microphone amplifiers

A MICROPHONE produces only a small AF signal, so you will invariably find the microphone connected to an AF amplifier in a transmitter block diagram. You saw this in the Foundation model.

An AF amplifier amplifies the small AF signal from the microphone to a level that we can use to modulate the carrier. It normally includes a filter to limit the AF bandwidth to a maximum of 3kHz, the minimum needed for voice communication.

Mixing frequencies (4c.1 & 4d.1)

WHEN TWO FREQUENCIES are mixed together, new frequencies are produced by adding the two original frequencies together and by taking one from the other. These are generally referred to as the 'sum and difference' frequencies.

Where one of the frequencies is an AF signal and the other is an RF signal, the mixing products are known as 'sidebands' and the sum and difference frequencies as the 'upper' and 'lower' sidebands.

CW transmitter

THE SIMPLEST type of amateur transmitter, shown in **Fig 44**, is a purpose-built CW set. It has no microphone, but a Morse key is used to 'modulate' the carrier by switching the signal path on and off in the 'keying stage'.

Fig 43: The circuit symbol for a microphone.

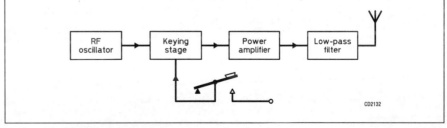

Fig 44: Block diagram of a CW transmitter.

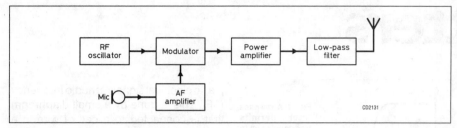

Fig 45: Block diagram of an AM transmitter.

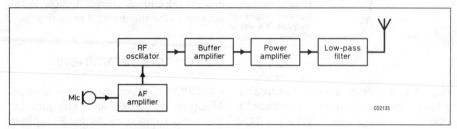

Fig 46: Block diagram of an FM transmitter.

The keying stage is the defining characteristic of a CW transmitter. All the other blocks have the same function as any other transmitter.

You may be asking 'why not key the oscillator or the power amplifier?' The simple answer is that we could, but such designs have been found to cause problems. If the oscillator is keyed it is like switching it on and off, causing changes in the DC supply and hence changing its frequency. This produces a bird-like CW note known as 'chirp'.

If we keyed the power amplifier we could have some quite high voltages across the key, causing sparks and interference known as 'key clicks'. The keying stage allows the RF oscillator and the power amplifier to run continuously and helps to prevent these problems.

AM transmitter (4d.2)

AM SIGNALS VARY in amplitude, in response to the AF signals from a microphone. The Intermediate course block diagram, shown in **Fig 45**, only differs from the Foundation model by the addition of the low pass filter.

However, you now need to know that the AM modulator produces an output that includes the carrier and two sidebands. The two sidebands are 'mirror images' of each other, both containing the same information. The carrier and both sidebands are amplified by the power amplifier and transmitted.

Let's look at an example. If an RF signal of 7050kHz were modulated with an AF signal of 3kHz, the output of the modulator would contain signals at 7050khz (the carrier), 7047kHz (the lower sideband) and 7053kHz (the upper sideband). The bandwidth of this signal (i.e. the difference between its highest and lowest frequency) would be 6kHz.

As we know, the AF bandwidth required for voice communications is 3kHz so the AF amplifier will normally include a filter that limits the signals from the microphone to a maximum of 3kHz. From this we can see that the bandwidth of an AM transmission should be 6kHz, twice the AF bandwidth.

FM transmitter (4d.6)

A FREQUENCY modulated (FM) transmitter works in quite a different way to an AM transmitter. As you can see in **Fig 46**, it modulates the RF by changing the frequency of the RF oscillator by a small amount. One way of achieving this is to connect the AF amplifier to the RF oscillator via a special diode known as a variable capacitance or 'varicap' diode. See **Fig 47** for a simplified circuit diagram.

A varicap diode acts like a small variable capacitor, with the capacitance being controlled by the voltage applied to it. As AF is an AC voltage, the capacitance varies in response to the AF signal applied to it. As the capacitance changes, the frequency of the tuned circuit that sets the frequency of the RF oscillator also changes in response to the AF signal. The amount that the frequency changes is known as the 'deviation'. In an amateur transmitter the deviation is limited to plus and minus 3kHz, giving a total bandwidth in the oder of 6kHz. FM channels must therefore be well separated to prevent adjacent channel interference.

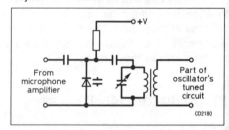

Fig 47: AF fed to a varicap diode in an oscillator.

Buffer amplifiers

The FM transmitter shown in **Fig 46** above includes a buffer amplifier. The purpose of a buffer amplifier is to isolate the RF oscillator from the rest of the transmitter circuits. This arrangement prevents any changes in the transmitter from changing the oscillator frequency. You may come across buffer amplifiers in any type of transmitter and some receivers.

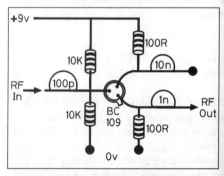

Fig 47a: Buffer amplifier circuit that can be added to the VFO on page 34 for use in practical circuits. Uses same 'ugly' technique - only ground two bottom resistors and 10n capacitor.

SSB transmitter (4d.3, 4d.4 & 4e.5)

THE MOST COMPLEX transmitter you will meet at this level is the single sideband (SSB) transmitter. A block diagram is shown in **Fig 48**. SSB is a form of AM, but a special modulator known as a 'balanced modulator' is used and a 'sideband' filter is added.

A balanced mixer mixes the AF signal with the carrier, but unlike the conventional AM modulator it only allows the two sidebands to pass. It removes, or at least suppresses, the carrier.

A sideband filter is a form of bandpass filter (**Fig 49**) that only allows a very narrow band of frequencies through; so narrow that it allows just one of the two sidebands to pass through whilst stopping the other. As you can see in **Fig 50**, this means that only one sideband is passed to the power amplifier, amplified and transmitted. The question usually asked at this point is 'Why remove the carrier and one sideband?' There are two main reasons.

1. First of all it means that you do not waste power transmitting a carrier that contains no information. Also, many years ago it was discovered that whilst the two sidebands of an AM transmitter contain the same information, only

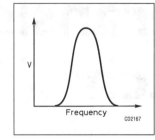

Fig 49: The characteristic of a bandpass filter.

one was required for effective communication. SSB is therefore far more efficient than AM or FM.

2. The second reason is associated with bandwidth. If only one sideband is transmitted the bandwidth of the signal will be only half of the equivalent AM signal, i.e. around 3kHz. This effectively means that around twice as many signals can fit into the same space on the amateur bands.

Data transmissions (4d.5)

MANY DATA transmissions can be made by taking audio signals from a personal computer and feeding them into the AF amplifier of your transmitter. Types of transmission such as Radio Teletype (RTTY), PSK31 and Slow Scan Television (SSTV) use one or two AF tones to modulate the carrier. These types of transmission can be used with SSB or FM transmitters. There are no specific block diagrams for data transmitters.

More information

IF YOU WANT to learn more about transmitters, either as part of your Intermediate training or after the assessments, the RSGB book *Practical Transmitters for Novices* explains more about oscillators, amplifiers and filters, as well as describing several amateur transmitters that you might like to build.

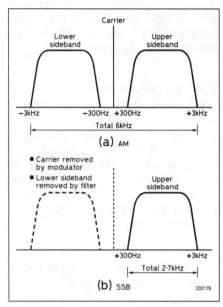

Fig 50: Bandwidth of DSB and SSB.

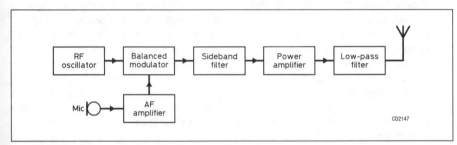

Fig 48: Block diagram of an SSB transmitter.

Using a Transistor as a Switch

TRANSISTORS HAVE many uses in amateur radio. In the following exercise you are going to use a transistor as a switch, using yourself as a conductor!

You will need your modified DC circuit (that is with the LED fitted, not the bulb) and an NPN transistor of any type. A plastic cased BC548 or metal cased BC108 are both suitable. The diagram of the pin-outs of these types are shown in **Fig 51**, but remember that these views of the pins are looking at them from *underneath*.

The exercise
(3h.9 & 10d.6)

1. First of all, place two of your fingers across the switch BC. Unless you have some very special properties, the LED should not light. Is that because you are not conducting any current through your body? We'll come back to that question later.

2. Identify the pins of the transistor from Fig 51 and then referring to **Fig 52** solder the collector (c) to the positive side of the switch B-C, at point C.

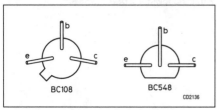

Fig 51: Pin-outs for BC108 and BC548 transistors, as seen from underneath.

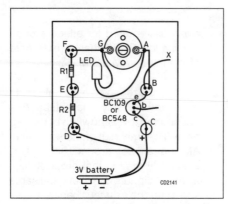

Fig 52: Layout of modification to the DC circuit.

3. Solder the emitter (e) to point B. Leave the base (b) free. Note that the LED still does not light up, because no current is flowing. The transistor does not conduct, because there is no 'turn on' voltage across the base and emitter.

4. Wet the end of your forefinger and middle finger, then place one on the positive point C and the other on the transistor base. **Fig 53** illustrates the circuit, and you are part of it! The LED should now light, indicating that a current is passing through the transistor. How can this be?

5. Try using one hand on positive point C and the other hand on the base. What effect does this have?

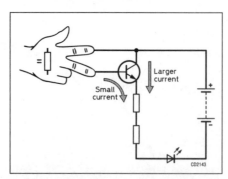

Fig 53: How the small current through your fingers causes a larger current to flow and light the LED.

Some answers

WHY DID YOUR fingers not light the LED on their own, when they did with the help of the transistor? Up to now you may not have believed that your body is a conductor, but clearly it is. However, your resistance is quite high, so there was not enough current flowing through your fingers to light the LED.

If you are still not convinced about the body being a conductor, switch your multimeter to read MΩ and hold one probe in each hand. Now measure the resistance between the fingers of one hand. Hopefully this will end any doubt.

The transistor on its own was not conducting, but by placing your fingers between the battery and the base you caused a small current to flow between the base and the emitter. This in turn caused a larger current to flow between the collector and the emitter, lighting the LED.

Using more of your body (hands, arms and chest) may have prevented enough current from flowing to light the LED, or it may have reduced its brightness. The result of that test depends on your resistance and that can be affected by how sweaty your hands are, the temperature and a number of other factors. Try it again after licking your fingers.

Summary

THE KEY POINTS to remember from this exercise are:

- The human body is a conductor – that is one of the reasons that electricity is so dangerous to us!
- Transistors will not conduct until a small current flows through the base to the emitter
- A small current through the base to the emitter causes a larger current to flow through the collector to the emitter – the transistor can therefore be used as a switch.

Licence Conditions 2

THERE ARE A FEW more items that you need to know at this level from the 'Terms, Conditions and Limitations', as the OfCom BR68/I booklet is called. Again, the more important ones have been picked out for the exam. As in the previous Licence Conditions worksheet, the style is to give the **syllabus item**, *the text from the licence document* and a few words of explanation.

2d.1 Recall that the licence holder may conduct Unattended Operation of a beacon, for the purposes of direction finding competitions, remote control of the main Station or for digital communications.

2(4)   Subject to sub-clause 2(5) the Licensee may conduct the Unattended Operations ("Unattended Operations" means the operation of the Station when unattended by the Licensee) only:
(a)   of a beacon
(b)   of a low power device to control apparatus at the Main Station address or a Temporary Location by remote control with a maximum power of 10mW erp pep.
(c)   by digital communications at the Main Station Address or at a Temporary Location:

Unattended Operation is one of the new privileges that your Intermediate Licence will give you. If you take a closer look at BR68/I you will notice that not all of the licence text has been copied here. That is deliberate, only the parts in the syllabus are shown here. If you intend to carry out any of these activities you really must read the full set of conditions, but for exam purposes it is quite sufficient for you to have a reasonable idea of what you can do and know where to check for the full details.

First of all you may operate a beacon. That is a device which transmits its callsign, usually in Morse, but it can be in synthesised speech, either all the time or on a prearranged clock cycle. Its purpose is to aid propagation research and allow amateurs to monitor changing band conditions. You may have heard beacons from other countries already. Details of frequencies on which you may find existing beacons are given in the *RSGB Yearbook*.

A slightly different kind of beacon is the use of a hidden transmitter used for a direction finding competition. If you are intending to do this you should first consult a more experienced amateur or join in one of the organised events that take place throughout the year.

Secondly, you may decide to locate your transmitter in a garden shed (strong and locked, hopefully) and control the equipment from the house via a very low power radio link. In licence terms the Station is now 'Unattended' and the clause in the licence allows you to do this. Some amateurs have computer-controlled transceivers. If the controlling computer is not in the same room as the transmitter, then that is also 'unattended', but the 'Unattended Operation' clause in the Licence does not apply.

The third type of Unattended Operation is a Packet station that can run for 24 hours a day, seven days a week. Most of the Terminal Node Controllers (TNCs) have the capability of recognising their callsign and storing an incoming message. They are also able to retransmit the messages if the station for which a message is stored happens to log on. Either way, the protocols (procedures) of the Packet system require each packet of data or message received to be acknowledged, even if the operator is not present. That means the station must transmit whilst unattended and the Intermediate licence allows that.

2d.2 Recall that 7 days notice of Unattended Operations must be given to the local office of OfCom

2(5)   The licensee shall not conduct unattended operation of a beacon or of digital communications unless he has given at least 7 days' written notice of the location, period of operation, frequency, power (Watts), identity of other users of wireless telegraphy who share the site and shut down procedure to the Operations Manager of the local office of OfCom in whose area the operation is to take place. The Operations Manager may, before the commencement of operation, prohibit unattended operation or allow the operation on compliance with conditions which he may specify.

Put simply, this means that if you intend to use your Station for Unattended Operation other than a low power remote control link, you must let the local OfCom office know and give them the chance to comment.

There are two key reasons for this. Firstly, if a fault develops and your transmitter starts to cause interference, a method of getting it switched off reasonably quickly has to be agreed. Secondly, many amateur bands are shared and, unknown to you, there may be another user, perhaps the Primary user, asking you to close down or change frequency. Again, OfCom need to know how to contact the Licence holder should this occur.

Basic transmitting log book, available from the RSGB.

2e.1 Recall that the Log may be written in a book (not loose leaf), on a magnetic disk or tape or other electronic storage medium.

2e.2 Recall that if the Log is kept on a computer, the means to view the Log and print a copy must be readily available and that suitable precautions must be taken to ensure the Log is backed up.

2e.3 Recall that the Log must be retained for at least six months after the last entry.

6(1) The Licensee shall keep a permanent record (the "Log") of all wireless telegraphy transmissions at the Main Station Address and all Temporary Locations showing:
(a) dates of transmission;
(b) the times (in Co-ordinated Universal Time (UTC)) during each day of
(i) the first and last transmissions from the Station (except when using automatic operations involving digital communications); or
(ii) switching the Station on and off for the purposes of enabling transmissions (when using automatic operations involving digital communications), changing the frequency band, class of emission or power;
(c) frequency band of transmission or, in an Unattended Operation, the specific frequency employed;
(d) class of emission;
(e) power;
(f) initial calls ("CQ" calls) (whether or not they are answered);
(g) except during automatic operations involving digital communications, the callsign of licensed amateurs or licensed stations with which communications have been established (not including those amateurs or stations which form part of the intermediate relay of Messages);
(h) details of tests carried out in accordance with sub-clause 4(4); and
(i) location when the station is operated at a Temporary Location.

6(2) The Log shall be written in a book or maintained on a magnetic tape, disc or other electronic storage medium. If the Log is maintained on an electronic storage medium the means to view the Log and produce a hard copy shall be kept readily available at the Main Station Address.

6(3) Where the Log is maintained:
(a) in a book, the book shall not be loose-leaf and no gaps shall be left between the entries;
(b) on a magnetic tape, disc or other electronic storage medium, suitable precautions shall be taken to ensure that the log is backed up.

What you must log was covered in your Foundation training and exam. At this level the need to be able to view and produce a printed copy of electronic logs has been added. If you decide to keep your Log electronically, the onus is on you to take suitable precautions to be able to view, print and back up your Log. Copying the Logbook data onto a floppy disk after each session would meet this requirement.

Another term has also been added to the items to be logged. Since an Intermediate licensee is at liberty to build transmitters, and not just use ready compiled kits, tests must be carried out from time to time to check that the station is operating as intended.

You need to record details of when you check out your Station in your Log. The actual checks are covered below and a later worksheet explains how to conduct the checks.

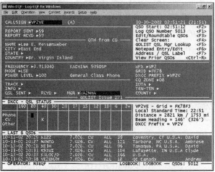

Screen dump of a computer logging program.

2f.1 Recall that transmissions from the Station must not cause undue interference to other radio users.

2f.2 Recall that the licensee must reduce any emissions causing interference, to the satisfaction of an Officer authorised by the Secretary of State.

2f.3 Recall that the licensee must carry out tests from time to time to ensure that the Station is not causing undue interference.

4(2) Notwithstanding any other term of this Licence, the Licensee shall ensure that the apparatus comprised in the Station is designed and constructed, and maintained and used, so that its use does not cause undue interference to any wireless telegraphy.
4(3) If any undue interference to wireless telegraphy is caused by the radiation of Unwanted Emissions or the field strength of electromagnetic energy radiated from the Station, then the Licensee shall suppress the Unwanted Emissions or reduce the level of the field strength to the degree satisfactory to the Secretary of State.
4(4) The Licensee shall conduct tests from time to time to ensure that the requirements of this clause 4 are met.

These licence conditions are all about electromagnetic compatibility (EMC) and good design of your transmitter. Other worksheets will help you understand the technical issues, but here we are concerned with the licence rules and what you may be asked in the exam.

The first point simply says you must not cause *undue* interference to other radio users. You may well think it would also be a good idea not to cause *any* interference to other electronic equipment such as a music system. In practice you are right, but the licence simply says "*wireless telegraphy*" so a CD player, for example, would not be covered. For the exam, "*wireless telegraphy*" is the right answer.

A little care is needed over the rather old-fashioned term "wireless telegraphy". This is a broad term for any radio system, be it real telegraphy (e.g. Morse or teleprinter), telephony (speech), other data, television, radar or even the radio signals from the GPS (global positioning system) satellites.

The licence refers to two key causes of interference:

- Unwanted Emissions. Any signals that the transmitter is producing that are not needed for communication and should not be there.
- Excessive field strength. That is the strength of your transmitted signal is too great for a reasonable piece of equipment to withstand without malfunction.

In either case Ofcom can ask (or tell) you to cure the problem to a level that does not cause undue interference.

2g.1 Recall the licensing role of Ofcom on behalf of the Secretary of State.

2g.2 Recall that possession of a current Validation Document is necessary for the Station to be used.

2g.3 Recall that if a licence is not renewed, it expires at the end of the day before the anniversary of the date of issue.

2g.4 Recall that the licence can be revoked by the Secretary of State.

9(2) *The Licensee shall pay to the Secretary of State before the anniversary date of the Date of Issue in each year, the fee on renewal prescribed by the regulations for the time being in force under sub-section 2(1) of the Act, and on payment of the fee the Secretary of State will issue to the Licensee a document in the form of the title page to this Licence (the "Validation Document") which will indicate the next date for renewal.*

9(3) *If the Licensee does not pay any fee due and in the manner described in sub-clause 9(2), then the Licence shall expire at the end of the day before the relevant anniversary date of the Date of Issue.*

All amateur radio licences are issued by the Secretary of State for Trade and Industry. Ofcom act on behalf of the Secretary of State and they sub-contract the work to the Radio Licensing Centre. Although your main contact will be with the Licensing Centre, you need to know how the others fit in.

Your Foundation Licence, and your Intermediate Licence when you get it, is valid for one year. At the end of that time you must renew the Licence or it will lapse. When you renew the licence you will be sent a Validation Document, which gives your Licence another year's life. You must keep these documents together.

You must pay any fee due before expiry of the current licence and if you forget, the licence expires at 23.59 hours on the day before you received it in the first year. If it was issued on 10 August, then it is valid up till midnight on 9 August the following year.

The final point is that the licence can be revoked. That does not normally happen unless you have persistently and deliberately ignored the terms of your licence and any warnings. Ofcom would very much prefer it if you stick to the rules and they will help as much as they can, especially if any difficulty is not really your fault. However, if you are not prepared to co-operate, you place your licence in jeopardy and could also find yourself in Court on criminal charges.

2g.5 Recall that the full licence conditions are set out in the Terms, Provisions and Limitations Booklet BR68/I

Perhaps rather obvious. You are only expected to know the key points about the terms and conditions of the Licence. However you must know when and how to check the full set of rules as set out in BR68/I.

2h.1 Be able to interpret the schedule to the Intermediate Licence.

The schedule, but not the full BR68/I, will be available in the exam. It is likely that a couple of the questions will be about the schedule. You will be expected to look up the answer. This is much the same as for your Foundation exam, but the questions might be slightly less straightforward. For example, the question may be 'What is the lowest frequency on which you transmit fast scan television (i.e. moving images like commercial TV)?'. An inspection of column 4 will show this to be 430MHz.

Remember that the Intermediate Licence allows additional privileges over and above the Foundation Licence, so the BR68/I Schedule includes more detail that BR68/F. You would be well advised to get used to reading the Intermediate Licence Schedule before the exam, but there's no need to try to memorise it all!

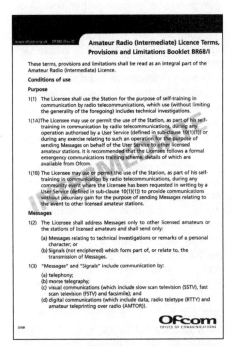

The front cover of BR68/I, the Terms and Limitations booklet for the Intermediate Licence.

Power Supplies

PRACTICALLY ALL amateur radio equipment needs some form of electrical power supply. Most modern transmitters, receivers and transceivers use a 12-14V DC supply.

Power can be supplied from a variety of sources including solar panels that convert sunlight and wind-powered generators that convert wind energy into electricity, but here we will focus on batteries and the domestic mains supply.

Typical mains transformers.

Batteries (3c.1 & 3c.2)

A PRIMARY CELL stores electrical energy by chemical means. When it is connected into a circuit, chemical energy is changed into electrical energy and DC flows.

Current will continue to flow until the chemical energy is exhausted. The cell is then said to be 'discharged' or 'flat'. Once a cell is discharged it must be disposed of in a suitable manner (if you need advice, contact your Local Authority Waste Disposal Office).

Primary cells can be used as a convenient and portable power supply for a low power station. However, because they can only be used once, they can prove to be expensive if replacements are required on a regular basis.

Secondary cells are rechargeable and use chemicals that can be discharged and recharged many times. Some of the chemicals found in rechargeable cells include Nickel and Cadmium (NiCad) and Lithium Metal Hydride (LiMH). Whilst these cells are more expensive to buy in the first place, they can be both convenient and cost effective in the long term.

If two or more cells are connected together, they form a 'battery'. This term has come to be used for cells and batteries, but you should remember that a single cell will generally only produce about 1.5 volts, whilst batteries are required for higher voltages.

Generally speaking, chemical cells and batteries can only supply low current, so they are limited to low power (QRP) operation. Lead-acid batteries, as used in motorcars, are another type of rechargeable battery. They are capable of producing much more current than conventional batteries and therefore make good power supplies for higher powered stations. However, there are some drawbacks. Not only are lead-acid batteries very heavy, they also contain acid that can attack carpets, clothing and skin! Great care is needed when carrying these batteries, in order not to hurt your back, fingers or toes from the weight, and to prevent the acid leaking out. The risks from acid spills can be avoided by using sealed gell-cell batteries.

Another thing to remember about lead-acid batteries is that the high current they can supply is also capable of generating lots of heat in a low resistance circuit. A wire short-circuiting the terminals of a lead-acid battery is likely to melt or even explode!

Mains power supply units

BATTERIES CAN provide excellent service for a mobile or portable station but at home you will probably want to take advantage of domestic mains electricity, which is relatively cheap and doesn't tend to run out in mid contact! However, you will remember that UK mains electricity is supplied at 230V AC, whereas our equipment tends to require a 12V DC supply. We therefore need some form of device to reduce the voltage and to convert the AC to DC.

Transformers (3g.1, 3g.2 & 3g.3)

TRANSFORMERS USE electro-magnetic properties to pass AC from one circuit to another without direct connection. They are sometimes referred to as 'isolating transformers'. Transformers are normally constructed from two quite separate coils of wire wound on the same former. The two coils are generally insulated from each other and are known as the primary and the secondary. The circuit symbol is shown in **Fig 54**.

The voltage provided by the secondary winding depends on the number of turns in each winding. If the secondary has half the number of turns of the primary, its output voltage will be half that of the primary. Similarly, if the secondary has three times as many turns, the output voltage will be three times that of the primary.

The coils in a transformer may be wound on an iron former to concentrate the magnetic field and prevent it affecting other components nearby.

Just as the current in a coil produces a magnetic field around it, moving a coil through a magnetic field will induce a current in the coil. This means that transformers only work with AC. AC produces a constantly changing magnetic field, giving the same effect as moving a coil through a magnetic field. Passing DC through the primary will *not* produce DC in the secondary, because the magnetic field is not changing or moving.

If the secondary has less turns than the primary, a smaller voltage will be

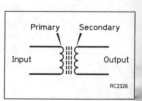
Fig 54: The circuit symbol for a transformer.

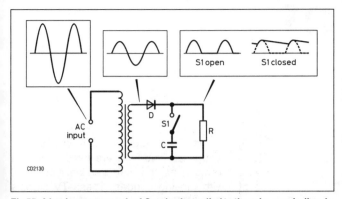

Fig 55: A basic power supply. AC mains is applied to the primary winding. In this instance a step-down transformer is being used, so a smaller AC voltage is developed in the secondary winding. This is rectified by the diode into pulsed DC, but if we close the switch the capacitor is brought into circuit and it smooths the voltage, although a certain amount of ripple remains. A practical power supply would not include a switch.

produced in the secondary. This is known as a 'step down' transformer. If the secondary has more turns than the primary, a higher voltage will be produced in the secondary. This is known as a 'step up' transformer. Both types of transformer have the same circuit symbol.

Modern amateur radio mains power supply units normally use step-down transformers to reduce the 230V AC from the mains to about 12V AC. However, there are still some transceivers in operation that use thermionic valves. The power supplies for valve radios often use step-up transformers to produce the hundreds - or even thousands - of volts required. These units can be *very* dangerous and the utmost care must be taken if using this type of equipment.

Diodes (3h.2 & 3h.3)

AS PREVIOUSLY STATED, a diode will only conduct in one direction. Therefore we can use a diode in our mains power supply - illustrated in **Fig 55** - to 'rectify' the 12V AC from the step-down transformer to produce DC. Note that you may be asked to recognise the various AC and DC waveforms shown.

A diode connected to the output of a step-down transformer will conduct whilst the AC is positive, but not when it is negative. This will result in pulses of DC varying from 0 to 12V. In practice you might use a number of diodes or a device known as a 'bridge rectifier' to rectify the negative parts of the cycle as well, but don't worry too much about that at this level.

Smoothing the supply (3h.4)

The pulses of DC are not suitable to power our equipment - we need a smoother supply. You will recall that a capacitor is able to store an electrical charge, so if we connect a large value capacitor to the output of the rectifier it will tend to charge up as the DC potential rises and discharge as the DC potential falls. In effect, a smoothing capacitor fills in the gaps between the DC pulses during the non-conducting part of the cycle.

Practical power supply units

THE DESCRIPTION in this worksheet contains a very basic outline of a mains power supply unit. This should not be used to build such a unit. A number of tried and tested designs for practical projects can be found in the RSGB *Radiocommunication Handbook*. You could consider building one of these for your assessment project. However, extra care is needed when building devices for direct connection to the mains, because of the high voltages and safety implications involved. If you are building a mains power supply or any mains device, you really should have it checked by a competent person.

The kind of smoothing capacitor you might find in a large power supply.

Other Types of Transmission

THERE ARE MODES apart from Morse and voice that Intermediate radio amateurs can use. We have already mentioned some, including Packet, RTTY and PSK31. These are generally referred to as 'data modes'. Let's have a quick look at each one in turn.

Data modes (8e.1)

PACKET RADIO is a system based on similar principles to those used in the Internet. It is commonly used to send and receive messages that are stored by other amateurs for later transmission through a network of linked stations. Although most Packet work is short range, using FM transceivers on VHF and UHF, the stations are connected in such a way that messages can be sent and received all over the world.

To set up stations for Packet, most amateurs use a Terminal Node Controller (TNC). These are specifically designed for amateur radio use. A TNC is a type of modem, which can be connected to the sound input and output of an FM transceiver (for example, at the microphone and loudspeaker sockets).

RTTY means 'radio teletype' and this is a system of data transmission that has been in use on HF for many decades. It is still frequently heard in amateur radio, although many operators think that PSK31 will gradually replace RTTY.

The abbreviation PSK31 refers to the method of signal generation (phase shift keying), and to the bandwidth of the signal generated, about 31Hz. This is considerably less than RTTY, which means that many PSK31 signals can use a

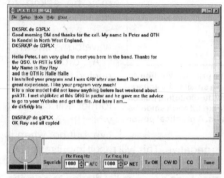
A typical PSK31 display.

smaller slice of the bands available to amateurs. PSK31 has been designed for keyboard-to-keyboard communication, and can still be effective even when signals are very weak (S1).

Amateur television (8e.1)

THERE ARE TWO main types of television transmissions that can be used by radio amateurs. Both break down images into lines and turn the lines into a codified sound for transmission. The sounds from the receiver are then decoded back into lines and the picture is 'reassembled'. The key difference between the two modes is the speed at which the lines are 'scanned'.

Slow Scan Television (SSTV) can be compared to a telephone facsimile (or fax) system in that it can only deal with 'still' pictures. It is a relatively slow method of transmission, but uses about the same bandwidth as a normal voice channel. Because of this, it can be used on the HF bands for the exchange of pictures across the world.

A typical RTTY screen.

A Slow Scan Television (SSTV) picture, transmitted from the Russian Space Station Mir.

Fast Scan Television (FSTV) is similar to the television we are familiar with at home in that the scan is so fast it can deal with moving images. These TV signals use a large bandwidth, much greater than SSB or even AM. They are therefore only suitable for use on UHF, where the amateur bands are wider (have a look at your Intermediate schedule to see how wide). The range is short due to the frequencies used but quality can be very good, and there are some FSTV repeaters to extend the range.

Amateur radio and the PC (8e.2)

RADIO AMATEURS have devised programs for personal computers that can be used with many of the modes mentioned in this worksheet. Communication to and from the radio is usually made via the computer's sound card, and software is used to generate and decode the signals with correct format and timing. Many of the programs are available free of charge from the Internet.

Advantages of different modes (8d.1)

FROM THE POINT of view of operating on the air, Morse is generally considered to be the most effective. It is simple to create and read, and has the greatest range for a given transmit power. Data modes such as PSK31 are arguably just as efficient, but they require more complex equipment, including a computer.

As far as voice modes are concerned, SSB is very effective, with the best range and lowest bandwidth. FM generally has a shorter range, but will normally produce a higher quality signal. In noisy conditions, Morse and SSB, which use lower bandwidths, will produce higher readability. This is because the bandwidth of the receiver can be reduced to the minimum, which cuts out as much of the noise or interference as possible without affecting the wanted signal too greatly.

The different modes also have different EMC characteristics. These are covered in another worksheet.

Orbiting satellites (8g.1 & 8g.3)

SATELLITES CARRYING amateur radio transponders (like repeaters) orbit the earth at heights above 150km. These satellites are designed and built by radio amateurs and are generally launched from routine commercial or scientific research spacecraft.

Once launched, the satellites do not remain stationary but move in relation to the earth and are only above the horizon at certain times. It is important to realise that the satellites can only be used when they are above the horizon at both the sending *and* receiving stations. Essentially, the antennas at both stations must be able to 'see' the satellite.

The movement of the satellite in relation to the earth causes some frequency variation on the received signal known as 'Doppler shift'. This variation must be allowed for when selecting your operating frequencies.

Satellite modes (8g.2)

THERE ARE SEVERAL combinations of up- and down-link frequencies used by amateur satellites. These combinations are known as 'modes' and in each case the up-link and down-link frequencies are in different amateur bands. For example, mode A has a 145MHz up-link with reception of the down-link on 29MHz.

You should note that if you decide to transmit through these satellites your station *must* be able to receive on the frequency you intend to transmit as well as the down-link frequency for the mode in use.

Satellite modes should not be confused with the modes used for different types of transmission. Indeed, unlike terrestrial repeaters, satellites can be used with CW, SSB and FM.

Details of the different satellite modes are published by a number of amateur organisations and examination of the band plans will show where satellite signals may be found.

Limiting power (8g.4)

THE GENERAL GOOD practice of only using the minimum RF power necessary to maintain effective communication is even more important when using satellites because they have very limited power supplies.

Solar panels are used to convert the energy from the sun's light into electrical energy to recharge the onboard batteries, but excessive up-link power from your station will cause the satellite to use more energy than is necessary, resulting in wasteful and unfair use of the satellite's power.

As a rule of thumb, your down-link signal should be no stronger than the satellite beacon signal.

Further information

PROBABLY THE BEST source of information that covers all these different types of transmission is the RSGB *Amateur Radio Operating Manual*, but there are other books and interest groups that specialise in each of the different modes.

For example, if you are interested in satellite operating you should contact AMSAT-UK who will be more than willing to help you get going. The British Amateur Radio Teledata Group (BARTG) is the organisation to contact for help with data transmissions.

Useful Contact Information

BARTG
British Amateur Radio Teledata Group
c/o 9 Linden Road
Oak Park, Cullompton
Devon
EX15 1TE

Web: www.bartg.demon.co.uk

AMSAT-UK
Badgers, Letton Close
Blandford Forum
Dorset
DT11 7SS

Web: www.uk.amsat.org

Receivers

AS WITH TRANSMITTERS, the Foundation course used a very basic model to explain the workings of an amateur radio receiver. Just as we learned more about transmitters earlier in this book, we are now going to look at receivers in a little more depth.

We start with the same three blocks that were used in the Foundation model, then build-up to a more complex model that resembles more closely the modern commercial designs found in the real world.

Functions of the blocks (4f.2 & 4h.1)

THE THREE BLOCKS in the Foundation receiver model (see **Fig 56**) appear in almost all receivers and serve the same purposes:

1. The tuning and RF amplifier selects the wanted frequency from all the frequencies picked up by the aerial. It will comprise, or at the very least include, some form of tuned circuit. Not all tuning stages include an RF amplifier, as we will soon find out.

2. The detector recovers the transmitted information from the modulated received signal. It is more correctly referred to as a 'demodulator'. There are several types of demodulator, as you will soon discover.

3. The AF amplifier raises the amplitude of the recovered audio to a level that can be heard through a loudspeaker or headphones.

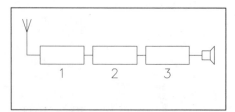

Fig 56: Foundation course receiver block diagram.

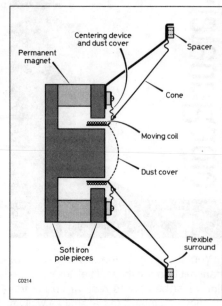

Fig 57: Cross section of a loudspeaker.

Loudspeakers (3i.1)

IN MANY WAYS the loudspeaker (block 4 in Fig 56) can be seen as the opposite of the microphone. The construction of a loudspeaker is illustrated in **Fig 57** and the circuit symbol is shown in **Fig 58**.

AC signals flowing through a coil of wire in a magnetic field cause a diaphragm, the speaker's cone, to move in response to the signals. The cone's vibrations cause mechanical movement in the air, which enables us to hear the sounds.

Fig 58: The circuit symbol for a loudspeaker.

The crystal diode receiver (4f.1 & 4i.1)

THIS TYPE OF receiver was one of the first ever designs, but it still works today. A circuit diagram can be seen in **Fig 59**. Quite remarkably, it doesn't need any batteries or external power supply to convert AM RF signals into audible AF sound waves.

The block diagram would contain blocks 1 and 2 from the Foundation model - the AF amplifier is omitted. In this configuration you need a suitable earpiece that will reproduce sound waves from the very low level AF signals.

It is possible to add an AF amplifier to drive a loudspeaker or standard earphones, but you will then need some form of power supply.

The tuning stage is generally formed by a large inductor with a variable capacitor connected across it. The detector is a single diode. There is no RF amplifier.

The diode demodulates the AM RF signal to produce a rectified signal that varies in amplitude, just as the audio signal at the transmitter did. The AF amplifier merely increases the level of the signals. These varying electrical signals cause the cone of a loudspeaker or the diaphragm of headphones to vibrate, reproducing the original sounds.

The crystal diode receiver has a number of limitations. It can really only detect quite strong signals, and it may tune in a number of stations at once. In other words it is not very sensitive or selective.

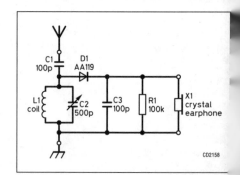

Fig 59: Circuit diagram of a crystal diode receiver.

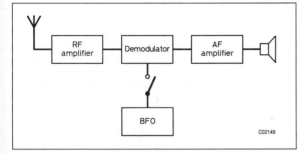

Fig 60: Block diagram of a TRF receiver.

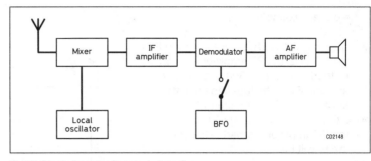

Fig 61: Block diagram of a superhet receiver.

The tuned RF receiver (4f.1 & 4f.2)

THE TUNED RADIO Frequency or TRF receiver (see **Fig 60**) also has three basic blocks, but it differs from the crystal diode receiver by including an amplifier in the tuning stage. You will also notice a block marked 'BFO'. We will come back to see what that is in a short while.

The amplifier increases the power of the RF signals selected by the tuned circuit before they are demodulated. This means that the receiver is much more sensitive than the crystal diode receiver. In other words, the receiver can detect much weaker signals, a distinct advantage.

Because the amplifier works at radio frequencies, it is known as an RF amplifier. Normally it includes an additional tuned circuit, making the TRF receiver far more selective than the crystal diode receiver. This means that it can tune to one station without picking up another station on an adjacent frequency as well.

Like the crystal diode receiver, the TRF uses a single diode demodulator and an AF amplifier to make the received signals audible. It is usually limited to AM use, although it can be used for other modes with some modifications.

The superhetrodyne receiver (4f.1 & 4f.2)

ALTHOUGH THE TRF receiver is much better than the crystal diode receiver, it still lacks sensitivity and is not selective enough for use on today's crowded amateur bands. The solution is not new, but is it effective.

The superheterodyne or 'superhet' receiver has been around since the 1920s, but it did not become really popular for many years after that. A block diagram of this type of receiver can be seen in **Fig 61**. Once again, ignore the BFO block at this stage, all will be revealed very soon! Today it is the basis of almost every receiver in commercial production, including television and broadcast radio receivers.

The block diagram of the superhet adds some new blocks to our main three ones:

- The Local Oscillator (LO) produces an RF signal in much the same way that the RF oscillator does in a transmitter
- The Mixer mixes or 'heterodynes' the selected RF signal from the antenna, or RF amplifier if fitted, with the RF signal from the local oscillator. This produces a new signal known as the Intermediate Frequency (IF).
- The IF amplifier. There is a limit to how much of an increase in sensitivity the RF amplifier can deliver - too much and it can start to oscillate. An IF amplifier is very similar to an RF amplifier, the basic difference being that it operates at a fixed frequency. It can therefore be made very selective and it adds more sensitivity, or gain, to the receiver.

Intermediate frequency (4g.1)

YOU MAY BE THINKING; 'Why bother mixing the RF signal with the local oscillator signal, just to make another signal at the IF?' Good question.

As you know, when we mix two frequencies together we get two new ones, the sum and the difference. You also know that we can use narrow filters to select the sum or the difference (as in an SSB transmitter). By selecting either the sum or the difference from the RF/LO mixer and amplifying it, we can make a very selective receiver with even more gain than the TRF.

By varying the local oscillator - using a VFO - we can tune across a band, but the output from the mixer will always be at the same frequency, the IF.

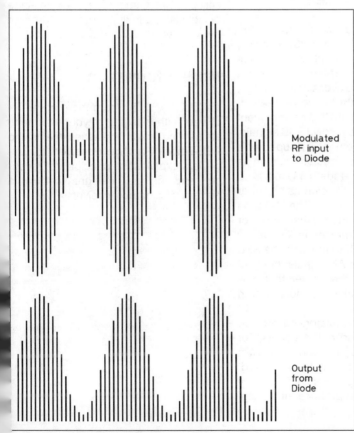

Fig 61a. Waveform produced by AM diode detector.
(You may be asked to recognise this in the exam)

Here's an example:

- Suppose the VFO can tune from 5000 to 5300kHz, the RF amplifier is tuned to the 80m band, 3500 to 3800kHz, and the IF is to be tuned to the difference (the VFO minus the RF).
- If the VFO is running at 5000kHz and the RF amplifier is tuned to 3500kHz, the IF will be:

    5000-3500 = 1500kHz

- If the same VFO is now shifted to 5300kHz and the RF amplifier is tuned to 3800kHz, the IF will be:

    5300-3800 = 1500kHz

Tuned circuits (4h.1)

FROM THE WORKSHEET on tuned circuits you will recall that tuned circuits can either select or reject specific frequencies. Both the RF and IF amplifiers will contain tuned circuits, comprising inductors and capacitors to select the wanted frequencies. The main difference between the two is that the RF amplifier will have to deal with a range of frequencies and must therefore include some form of variable capacitor or inductor, whereas the IF is always the same frequency so it does not need a variable tuned circuit.

It may not be obvious from the outside, but when you tune your superhet receiver you are actually shifting the local oscillator frequency and tuning the RF amplifier at the same time. This is known as 'tracking'. It makes construction slightly more complex, but operation is quite straightforward.

Different types of demodulator (4i.1)

THE TYPE OF SIGNAL at the output of the IF amplifier depends on the type of transmission. All that the receiver has done up to this point is to amplify the RF, change it to the IF and amplify it again. It matters not if the signal is CW, AM, FM or SSB.

Both the crystal diode and the TRF receivers used a single diode as the

Demodulators for different modes	
AM	single diode
FM	frequency discriminator
CW	diode with BFO
SSB/CW	product detector with CIO

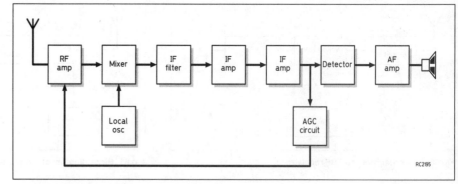

Fig 62: Block diagram of AGC.

demodulator. The superhet can use a single diode too, but the diode demodulator limits their use to AM signals. However, both TRF and superhet receivers can be used for CW reception by adding a new block, the Beat Frequency Oscillator (BFO).

The BFO is another type of oscillator. It runs at almost the same frequency as the RF in the TRF receiver, or the IF in the superhet receiver. By carefully tuning the BFO to be about 1kHz away from the signal at the detector, the two signals mix to produce sum and difference signals, one of which is at AF.

For example, if a BFO running at 1501kHz was mixed with the IF used earlier, the difference would be:

1501-1500 = 1kHz

Note that the sum would not be AF! (1501+1500=3001kHz=3.001Mhz) Audio frequencies around 1kHz are commonly used to listen to Morse code. BFOs are usually tuneable over a narrow range, to allow the tone of the Morse to be varied.

The diode detector/BFO arrangement can also be used to resolve SSB signals, but a better arrangement in a superhet is to use a special demodulator known as a Product Detector, together with a fixed frequency oscillator known as a Carrier Insertion Oscillator (CIO). The CIO runs slightly lower in frequency than the IF for USB and slightly higher in frequency than the IF for LSB, to mix with the SSB and reproduce the AF signals. In effect the CIO replaces the carrier that was removed by the balanced modulator in the transmitter.

None of the demodulators covered so far will properly resolve FM signals. For FM a special type of demodulator known as a Frequency Discriminator is required. The discriminator detects small frequency changes in the IF to produce the AF signals.

Automatic gain control (4i.2)

MANY RECEIVERS include Automatic Gain Control (AGC). This is a system designed to maintain a steady level of AF output, despite changes in the strength of the RF signal at the input of the receiver. There are several ways of achieving this, but one of the most common samples the output from the IF amplifier and feeds it back to the RF amplifier. It is illustrated in **Fig 62**.

As the output from the IF decreases, the lower AGC voltage causes the RF amplifier to increase its gain, making the weaker signals appear stronger. Once a stronger signal appears and the output from the IF increases, the higher AGC voltage reduces the gain in the RF amplifier. Once set up, the AGC should keep a wide range of signals at a fairly constant volume for the listener.

More information

IF YOU WANT TO learn more about receivers, either as part of your Intermediate training or after the assessments, the RSGB book *Practical Receivers for Beginners* explains the principles of reception and includes designs to build a number of amateur receivers.

Antenna Matching

THE ANTENNA AND its feeder are vital parts of an amateur radio station. The best transmitting and receiving equipment in the world will be almost useless if it is not connected to a good antenna and the best antenna in the world will be equally useless if the RF signals are not carried to and from your radio along an efficient feeder. What's more, an effective antenna system is something that you will have to create for yourself and what is best for you will depend to some degree on your own circumstances. Everyone's roof or garden is different, and your constraints are not the same as anyone else's.

Even if the rest of your station consists of commercial equipment, your antenna is more likely to be home made. Therefore it is important that you know some of the theory, so that you can put up the most effective. It will also be useful for the exam!

A little revision

DOUBTLESS YOU WILL remember from your Foundation training that one of the most commonly used antennas is the half-wave dipole (see **Fig 63**). This is more often referred to simply as a 'dipole'.

Radio amateurs probably use the dipole more than any other antenna on the HF bands. It is also the basis for many VHF and UHF antennas. After a few simple calculations, a dipole can be constructed easily. It is simple to feed and can be working in a matter of an hour or so.

As you will remember from your Foundation assessment, a dipole has to be adjusted or cut to the correct length. Because we are talking about a 'half wave' dipole, you would be correct in guessing that this antenna should be one half of the wavelength in use. In reality, factors such as the closeness of the ground affect the exact length required, so amateurs usually make the antenna too long to begin with, then cut equal lengths off each end until a good SWR reading is found. How does this work?

A transmitter needs a load (5e.1)

TRANSMITTERS ARE designed to work into a load. The antenna system (i.e. the antenna and its feeder) normally provides that load. If there is no load, or a load that is not what the transmitter is designed for, the power amplifier can be severely damaged.

To illustrate this point, think about pushing a trolley in the supermarket. If you are expecting a heavy load you will push hard, but if the load turns out to be very light the trolley will be pushed away and you may fall flat on your face. On the other hand, if the load turns out to be much heavier than you had expected, you could pull a muscle or injure your back. Either way, you could be 'damaged' by a load that you were not expecting.

Most modern amateur transmitters need to be connected to a load with an impedance of 50Ω. If the impedance varies too much from this value, the transmitter could be damaged. Some transmitters have an automatic power-reducing circuit to avoid damage caused by incorrect matching, but you should not rely on this.

What is impedance? (5b.1)

IF IMPEDANCE is measured in ohms (Ω), it must have something in common with resistance. This is true, but you cannot measure impedance with an ohmmeter as an ohmmeter uses a DC circuit and impedance only comes into play with AC. You may wish to look upon impedance as 'AC resistance'.

It is also true that a 50Ω resistor can be used as a load for a transmitter to work into, for example, as a dummy load for test purposes. However, in an antenna system, instead of the transmitter power being absorbed in the load, as in the dummy load, the power is transferred through the feeder to the antenna to produce a radio wave.

We don't have to investigate the theory in any detail here, it is simply sufficient to know that a properly tuned antenna system will only provide the correct load if we feed it from a transmitter at the correct frequency.

So what makes the antenna system impedance equal to 50Ω?

Feeder impedance (5b.2)

THE FIRST COMPONENT in the antenna system is the feeder.

The feeder is the cable that carries the RF signals between your radio and the antenna. Why don't we plug the antenna straight into the back of the radio, instead of using a feeder? The answer to this question is that the maximum signal, both on transmitting and receiving, is obtained by having the antenna in the open and as high as possible. This is also part of good EMC housekeeping.

Several types of feeder are used by radio amateurs and each type has a 'characteristic impedance'. As mentioned above, impedance is not the DC resistance of the wire but an AC characteristic. The characteristic impedance of coax is determined by the diameter and spacing of the conductors. Coaxial cables for amateur radio are normally 50Ω impedance, to suit modern transmitters. The coax used for television downleads is normally 75Ω and satellite TV coax tends to be 90Ω.

Generally speaking, if you buy a length of feeder, you should make sure it has the

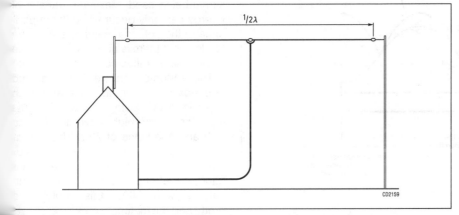

Fig 63: A dipole mounted horizontally in a garden.

characteristic impedance that you want. In most cases the length of the coax has no effect on the impedance – if there is a 50Ω load at one end, it will be 50Ω at the other end.

The next part of the system is the antenna itself, and antennas also have impedance.

Feedpoint impedance (5c.1)

ALL ANTENNAS have impedance or, more accurately, each antenna has a feedpoint impedance.

You should recall that a half-wave dipole has two sides or 'elements'. When it is cut to the correct length for the frequency in use, the impedance at the centre of the dipole is somewhere around 50Ω. Different antennas have different feedpoint impedances, but because the dipole is so common, most amateur antenna systems are calculated by reference to an impedance of 50Ω.

The feedpoint impedance of antennas is not fixed; it depends on a number of factors. There are equations to calculate feedpoint impedance from known values of potential difference and current, similar to the way we use Ohm's Law in DC circuits. You need to know that these relationships exist, but you don't need to know the equations and you certainly will not be asked to carry out any calculations using them.

Matching

WE NOW KNOW that modern transmitters need to work into a load of 50Ω, that commonly-used coaxial cable has a characteristic impedance of 50Ω, and we have discovered that a dipole has a feedpoint impedance of about 50Ω. If the transmitter, feeder and antenna all have the same impedance, we say that they are 'matched'.

The benefit of using a matched system is that no matter how long the feeder is, the impedance at the transmitter will always be the same as the impedance at the antenna, just as if the feeder were not there. In such circumstances the transmitter will not be damaged and we will have the most efficient transfer of energy from our transmitter to the antenna.

But what happens if there is a mismatch? There are two points where this can happen, between the transmitter and the feeder and at the junction of the feeder and antenna. Let's investigate this further, starting at the antenna.

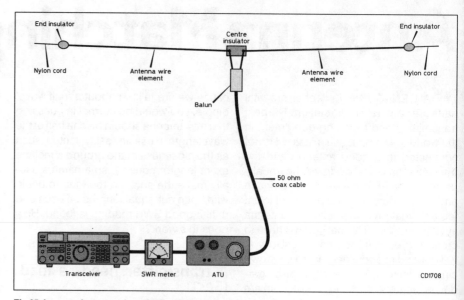

Fig 65: Layout of a transceiver, SWR, ATU, feeder, and antenna system.

Mismatch at the antenna (5c.1 & 5d.1)

IF AN ANTENNA is not adjusted to the correct length for the frequency in use, some of the RF power arriving from the transmitter will be reflected back down the feeder, as **Fig 64** shows. The reflected power is not lost, but it will combine with the power travelling up the cable to form what are called 'standing waves'. These are points of high and low voltage, and if you could see the pattern they form, it would look a little like stationary waves along the cable. You measured Standing Wave Ratio (SWR) as part of your Foundation assessment.

The reason that some of the power is reflected is that if the antenna is not the right length for the frequency in use, the feedpoint impedance will no longer be 50Ω. The antenna will only present the correct impedance when the antenna's length is correct for the wavelength of the signal being used. Maximum transfer of energy can only occur when the impedance of the antenna matches that of the feeder, so if there is any mismatch some power will be reflected.

The proportion of power that is reflected depends on how much of a mismatch there is. When using a feeder of 50Ω impedance, we might have an antenna with an impedance of 70Ω. This is reasonable, and would actually be considered a fairly good match. On the other hand, if the feedpoint impedance of an antenna were 500Ω, this would be a pretty bad mismatch.

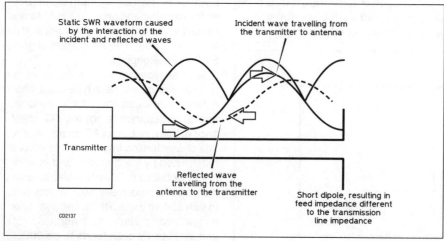

Fig 64: Standing waves on the feeder of an antenna.

The amount of power reflected is related to the size of the mismatch and the reflected power sets up the standing waves, so it is easy to visualise that the amount of standing waves depends on the size of mismatch. The greater the mismatch, the greater the standing waves. When we measure the SWR on a feeder we are, in effect, finding out to what extent our antenna system is mismatched. A lower SWR means a better match.

Mismatch at the transmitter (5d.1 & 5e.1)

THE MAIN PROBLEM of having a mismatched antenna is that the standing waves alter the impedance at the transmitter's antenna socket. Instead of the 50Ω of a well-matched system, the transmitter will instead be connected to some other impedance. We don't need to know how to calculate this, it is enough to say that the transmitter will not operate at full effectiveness if there is a mismatch between the transmitter and the feeder.

Most transmitters can tolerate a small amount of mismatch. This is why we can successfully use a single antenna over a whole amateur band. When we change frequency within the band the match between the antenna and feeder will change because the wavelength will be slightly longer or shorter, but the antenna stays the same length. As a result, the level of standing waves will change and the transmitter will 'see' different impedances at its antenna socket. However, because the mismatch will be quite small, the transmitter should be able to cope without difficulty.

Antenna matching units (5e.1)

WHAT HAPPENS if the change in frequency is much greater? On the HF bands it is common practice to use a single antenna on several different bands. Such an antenna will only be perfectly matched at one spot frequency and will normally only present an acceptable impedance across one band. On all the other bands the rather large mismatch will result in a high SWR. If this is the case, you can use an Antenna Tuning Unit (ATU) between the transmitter and feeder.

Fig 66: Make sure you use a ladder at the correct angle. Don't lean out!

An ATU can change the mismatched impedance that the antenna system presents to the transmitter to something more acceptable. This will allow the transmitter to feed power through the feeder into the antenna. This is why ATUs are sometimes called Antenna Matching Units (AMUs). The position in which an ATU would be used in your station is shown in **Fig 65**.

An ATU connected between the transmitter and the feeder will not remove the mismatch between the antenna and the feeder. Neither will it change the level of standing waves on the feeder. What it will do is let the transmitter 'see' the correct impedance and so feed the maximum power into the antenna system.

You should note that an ATU connected to a transmitter does *not* tune the antenna, so although you may have a 'perfect' SWR reading at the transmitter the antenna may not radiate very well. Remember that a dummy load presents an excellent match, but it is not a good radiator!

You should also be aware that when the transmitter output is correctly matched, an ATU can provide additional protection against the radiation of harmonics.

Safety note (9c)

THE QUEST FOR AN effective antenna system may require you to work at some height. The use of ladders and/or the lifting of long support poles are hazards that pose some risks. However, if you are aware of the risks, you can guard against them causing any real harm.

The first thing to be aware of is the presence of any overhead power lines. Any contact with high voltage cables will undoubtedly kill you. Even if you only get close to them, electricity can arc across the gap. *Never* work under or near overhead power cables and always think what might be hit if the ladders or poles were to fall!

The second thing to think about when working at height is to make sure that ladders are secured at the top and/or that you have another person standing on the bottom rung of the ladder. Younger amateurs should make sure that there is always an adult around when working at height.

As shown in **Fig 66**, all ladders should be set at the correct angle and you should not reach away from the ladder such that you cannot keep two feet firmly on the same rung.

If you need to take tools up a ladder, you may find a tool belt useful to prevent heavy objects falling onto your assistant(s) - who should be wearing suitable hard hats anyway! Tool belts and hard hats can be purchased at most good DIY retailers and builders' merchants.

Further information

IF YOU WOULD LIKE to learn more about antennas, there are many sources to refer to, but two RSGB books come highly recommended. *Practical Antennas for Novices* is a good starter, whilst *Backyard Antennas* takes things a bit further. Both books contain designs for building antennas, together with some of the technical background.

Antenna Feeders

YOU LEARNED IN your Foundation studies that coaxial cable is the most common type used by amateurs. In this worksheet we will reveal some more information about coax and introduce you to another common feeder; 'balanced' or 'twin'.

Coaxial cable
(5a.1, 5a.3 & 5b.2)

THE CONSTRUCTION of coax was covered at Foundation level and the photo below should refresh your memory.

You should recall that the braid ensures that the signal is contained within the cable, so that there is no chance that signals will be radiated from the feeder when transmitting. That said, be careful when connecting the antenna - this useful property may be reduced if you don't connect a balun between the unbalanced coax and a balanced antenna such as a dipole.

As well as preventing radiation from the feeder, the fact that all the RF signal is contained within the cable means that anything outside the cable will not affect the signals. Coax can be laid on the ground, routed under carpets or strapped to a metal post with no unwanted effects, either on transmitting or receiving.

You may have heard of coax being referred to as 50Ω cable. As you have now discovered, this is not the resistance of the wire but the characteristic impedance of coax and that this impedance is determined by the size and spacing of the conductors. You should note that it is not just the size of the coax that determines the impedance; 50Ω coax comes in several different sizes. However, most manufacturers mark the characteristic impedance on the outer jacket to prevent confusion.

Balanced feeder
(5a.1, 5a.2 & 5b.2)

ANOTHER COMMON TYPE of feeder used by amateurs is called 'balanced' or 'twin' feeder. There are several types of balanced feeder, but they all have two parallel wires. In some the spacing is quite wide, several centimetres, in others the wires are only separated by the insulation on them, a bit like hi-fi speaker cable. One of the more common types is often referred to as 'open-wire' feeder. This can be made at home at low cost and is very efficient.

How does balanced feeder work? That is, why doesn't a transmitted signal get radiated from the feeder? After all, there is no enclosing shield as there is with coaxial cable.

First of all you need to remember that as current flows along a conductor an electromagnetic field is formed around it. As RF is a form of AC the field around the conductor is constantly changing polarity.

In balanced feeder we have two conductors being fed by the same RF signal. **Fig 67** shows that as the AC goes positive on one conductor it goes negative on the other. The RF fields formed around the conductors are therefore equal and opposite and, because the wires are sufficiently close together the RF fields cancel each other out. This means that as long as balance is maintained, there will be no radiation from the feeder and all the RF energy will be fed to the antenna.

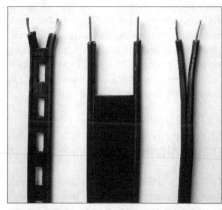

Balanced feeders. You may also come across balanced feeder that is made from two separate wires, held at a fixed distance by plastic 'spacers'.

However, as there is no braid or screen, with balanced feeder you must be careful not to let the feeder get too close to any other objects. It must not be buried or attached to walls or metal posts. Anything close to the feeder will disturb the balance and cause it to radiate.

Despite these limitations, there are benefits from using balanced feeder. For a start, it can be connected directly to a dipole, without the need for a balun. This is because we are connecting one balanced element to another. Using balanced feeder can also significantly reduce the possibility of EMC problems. However, a balun or a balanced ATU has to be used between the radio and the feeder, as most modern transmitters have unbalanced antenna sockets.

You may be asking: "What is the characteristic impedance of balanced feeder?" The same rules apply to balanced feeder as for coax. The characteristic impedance is determined by the diameter and spacing of the conductors. The types of balanced feeder that radio amateurs use generally have a characteristic impedances between 75 and 600Ω. You may already have encountered 300Ω ribbon feeder, which is the type often used for VHF FM broadcast antennas.

Coax cable with the layers gradually revealed. Good quality cable (as shown here) has a thick braid. In poor quality coax the braid is thin and wispy.

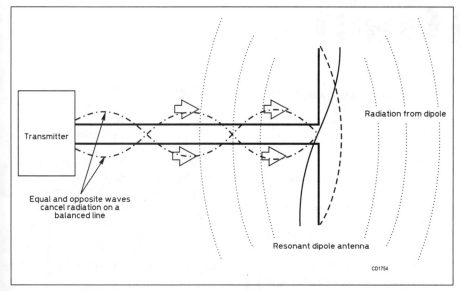

Fig 67: Equal and opposite currents cancelling in balanced feeder.

Feeder losses (5a.4 & 5a.5)

PROBABLY THE GREATEST benefit from using balanced feeder is concerned with losses.

All feeders cause some loss of signal strength - as the RF currents flow along the cables some of the energy is 'used up' by converting to heat within the cable. This applies to both transmitting and receiving, although we are usually most concerned about losses in our transmitted signal. As you might expect, the longer the feeder, the greater the loss.

The losses in balanced feeders are generally much lower than those experienced with coaxial cable. In fact, when used correctly, balanced feeder hundreds of metres long can be virtually loss-free. By contrast, coaxial cable suffers more loss.

Decibels in loss calculations (5a.6)

YOU NEED TO BE ABLE to estimate how much loss a feeder might cause to a signal. Manufacturers supply figures for the loss for different lengths of feeder under various circumstances. These will be quoted in decibels, which is usually abbreviated to 'dB'. The dB is a scientific way of expressing how much smaller (or bigger) a quantity will be compared to what it was to begin with.

You don't need to know how decibels are calculated at this stage, but you do need to be able to make use of the common decibel figures shown in **Table 5**.

The key facts to remember about decibels are:

- A loss of 3dB corresponds to a halving of power.
- Decibels can be added together.

Why is this important? Your licence conditions quote the maximum power you are permitted to use on the bands. This is normally the 'output' power at the antenna. If you have losses in your feeders you may not be making the most of your licence.

Loss (dB)	Fraction of power remaining
3	One half
6	One quarter
9	One eighth
10	One tenth

Table 5: Chart showing losses in dB and amount.

Examples

1. Your transmitter supplies 40 watts to the feeder. You see from the manufacturer's data that you will have a loss of 6dB in the feeder. How much power will reach the antenna?

   If you had a loss of 3dB, you would halve the power, and the power reaching the antenna would be 20 watts. With another 3dB of loss there is another halving of power, so the power reaching the antenna with 6dB of loss will be 10 watts.

2. You buy a new length of coax and feed it with a transmitter power of 20 watts. When you measure the output power at the end of the coax and find that it is only 2.5 watts. How much loss does the coax have?

   To calculate this, keep halving the power (i.e. step downwards by 3dB at a time) until you reach the test result. Half of 20 would be 10 watts (3dB), half of 10 would be 5 watts (another 3dB) and half or 5 is 2.5 (a third step of 3dB). The total loss is therefore 3 + 3 + 3 = 9dB.

3. Suppose the manufacturer tells you that the coax has a loss of 1dB for every 10m. How much loss would you have in a feeder that was 30m long?

   Again, the decibels are added together. Three lots of 1dB = 1+1+1 = 3dB. Feeding 50 watts into such a cable would only see 25 watts reaching the antenna.

It is worth mentioning that a loss of 10dB equates to a reduction to one tenth of the input. In the example above, 3dB loss reduced 50 to 25watts. If the total loss had been 10dB, the output would have been just 5 watts.

Antenna Gain

Decibels in gain calculations (5f.2)

YOU LEARNED AT Foundation level that some antennas have a property known as gain. An antenna with gain focuses RF energy, so it appears to be radiating more power than it was supplied with.

You should also recall the term 'Effective Radiated Power' (ERP). ERP is the power that appears to be transmitted from the antenna in the direction of maximum radiation as a result of the gain. To calculate the ERP we need to know the power supplied to the antenna and the gain of the antenna.

At Foundation level, gain was only referred to as 'a gain of 2 times' or 'a gain of 3 times', but at Intermediate level you need to know that antenna gain is expressed in decibels (dB). The dB is a scientific way of expressing how much bigger (or smaller) a quantity is, compared to what it was to begin with. As figures for gain are usually expressed in decibels we need to know how to use dB for gain calculations.

You don't need to know how decibels are calculated, but you do need to be able to make use of the common decibel figures shown in **Table 6**. When we are talking about gain, each step of 3dB represents a doubling of power. For example, suppose your transmitter supplies 30 watts to the antenna, which is a Yagi having a gain of 9dB. What is the ERP? In other words, what effective power is being radiated in the direction of maximum radiation?

A doubling of power to 60W would represent a gain of 3dB. Another doubling to 120W would represent 6dB, and further doubling to 240W adds another 3dB, making 9dB. So your ERP is 240W and your power in the wanted direction

Gain (dB)	Power multiplication
3	Two times
6	Four times
9	Eight times
10	Ten times

Table 6: Chart showing gain in dB and amount.

appears to have increased eight-fold (8 x 30 = 240).

It is worth mentioning 'the 10dB rule'. In the case of gain, 10dB represents a tenfold increase in power. If you feed 50 watts into an antenna with 10dB gain, your ERP will be 500 watts. Subject to the specific Schedule for the frequency in use, it is quite legal for an Intermediate Licence holder to do. Generally, the power quoted in the Schedule is the power supplied to the antenna, but there are some that limit the ERP. Remember to check the Schedule for the band you are using before spending money on a high gain antenna!

Polar diagrams (5f.1)

MOST ANTENNAS do not radiate equally in all directions, the main exception being the vertical ground plane antenna. The same applies to receiving - any gain that an antenna has works when receiving as well as transmitting. That is why you have to point your TV antenna towards the transmitter to get a good picture.

Many antenna systems used by amateurs are based on the half-wave dipole. This generally runs horizontally, with wires suspended from one or two supports in a back garden. When used in this way, the dipole is a directional antenna.

The maximum radiation from a dipole occurs from the sides, with minimum off the ends. This means that if your dipole runs north to south, its best directions will be to the east and west. If you want to know how great this effect is, look at a polar diagram such as **Fig 68**.

The polar diagram shows the antenna as a straight piece of wire and we imagine we are looking down on it from the sky. The bow-shaped curve on either side of the dipole is a representation of the radiation from the dipole. Lay a pencil from the centre of the antenna to any point on the curve: the distance from the centre to the curve itself represents the strength of radiation you can expect in the direction the pencil is pointing.

From this you will see that there is a fairly broad pattern of radiation out from the sides of the dipole, without too much change over quite a wide sweep. But there is a noticeable reduction as you move toward the ends of the antenna, with no radiation at all from either end. In real life, nearby objects will reduce this effect.

The Yagi antenna (5f.3)

WE TALK ABOUT gain when using an antenna such as a Yagi (**Fig 69**). To refresh your memory, you can think of a Yagi as being based on a half-wave dipole. In the case of a 3-element Yagi, the 'driven element' is in the centre and is attached to the feeder. The 'reflector' is slightly longer than the driven element and the 'director' is slightly shorter. Both are parallel to the driven element, and the direction of maximum radiation is towards the director – it directs the transmitted power.

The Yagi antenna is designed to maximise radiation in one direction. The polar diagram of a Yagi depends on many factors, including the number of elements and their spacing. The key factor of the Yagi is that there is far more radiation in one direction, with a small amount to the rear and very little off the sides. A typical polar diagram can be seen in **Fig 70**.

Practical exercises (5f.3 & 5f.5)

IF YOU HAVE a VHF dipole, take it out into the open with a portable receiver. Using the dipole horizontally (parallel to the ground), try rotating it whilst listening to a steady signal on the band. Observe the S meter and you should find that there is a

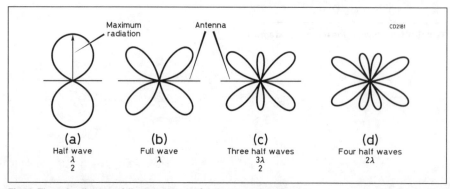

Fig 68: The polar diagram of dipole antennas of various wavelengths.
(Note that you are only expected to recognise the half-wave dipole in the exam)

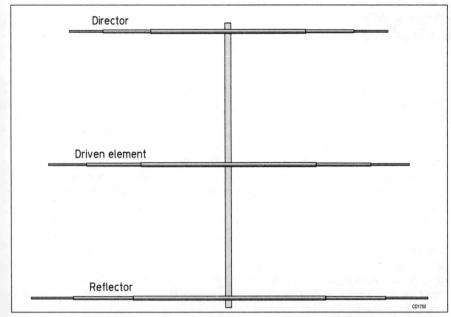

Fig 69: A three element Yagi antenna.

dip in the received signal strength when either end of the dipole is pointing at the transmitting station. The polar diagram shows the direction(s) for maximum reception as well as maximum radiation.

Another good demonstration is to try rotating a VHF or UHF Yagi whilst listening to a steady signal on the band. You should find that there is a peak in the received signal strength when the director is pointing at the transmitting station. This technique is used in the national and international direction finding competitions that many amateurs enjoy taking part in throughout the year.

Finally, you might like to see what effect changing the polarisation of the antenna has (i.e. using it horizontally or vertically). You should find that the received signal is much stronger when the receiving antenna is using the same polarisation as the transmitter.

Safety note (9e)

RF ENERGY can be hazardous to your health on two fronts. Firstly, as you learned at Foundation level, RF currents can cause severe contact burns and you should never touch an antenna whilst transmitting (except for a small handheld, perhaps).

The second safety point relates to the fact that RF waves can penetrate the human body. Current scientific knowledge suggests that the main hazard from this penetration is that body tissue will be heated as it absorbs the energy – the exact property we use in a microwave oven. However, the authorities maintain that an amateur radio station is unlikely to exceed the exposure limits they have set.

Despite these assurances you should avoid getting too close to high gain antennas and you should never look down the waveguides used for the microwave transmitters that you will be permitted to use with your Intermediate Licence. High gain antennas such as dishes focus RF into a more intense beam of energy and waveguides focus microwaves into an even smaller area, increasing the heating effects. Eyes are particularly at risk from heating.

If you are at all concerned about electromagnetic radiation safety, you can find out more from the National Radiological Protection Board (NRPB), the UK authority on the matter. International exposure levels have also been recommended by the World Health Organisation and ICNIRP.

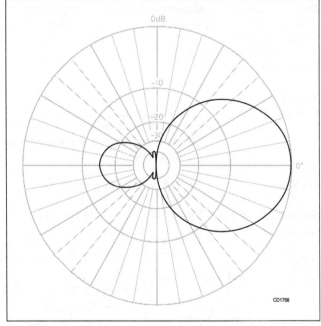

Fig 70: The polar diagram of a typical 3-element Yagi.

Propagation

PROPAGATION IS one of the most fascinating aspects of amateur radio. It is an aspect of the hobby that is still not fully understood, even by experts. However, there are some well-known general principles that you need to be aware of. Learning about propagation will give you an insight into what distances can be achieved and will help you get more out of the hobby.

Frequency and wavelength (6a.8)

AT FOUNDATION level you should have learnt that the frequency of an alternating signal is measured in Hertz (Hz). One Hz means one cycle per second. In amateur radio we often use the units kilohertz (kHz) and Megahertz (MHz). You need to remember that 'kilo' means a thousand and 'Mega' a million. A signal at 10MHz is therefore completing ten million cycles in one second.

There is a unit for even greater frequencies. 'Giga' (G) means one thousand million, so signals in the 10GHz band will be completing ten thousand million (10,000,000,000) cycles per second! This frequency could also be expressed as 10,000MHz, with the M representing the other six zeros.

If you find these units confusing, just remember to add three zeros for each step up:

- 1kHz = 1,000Hz
- 1MHz = 1,000,000Hz
- 1GHz = 1,000,000,000Hz

You may recall that the frequency of a radio wave is related to its wavelength. If you visualise the signal as a wave travelling through space, alternating a certain number of times per second (so many Hertz), the distance from one peak to the next of this wave - as shown in **Fig 71** - is one wavelength.

Wavelength is normally measured in metres. For the Foundation exam you were allowed to use a chart for converting frequency in MHz to wavelength in metres, and vice versa. You are now at a level where you can be asked to convert frequency to wavelength without a chart. Don't panic, it's not that difficult!

The link between frequency and wavelength is easy to understand if you remember that *all radio waves* travel at the same speed (velocity). So, if the wavelength of a radio wave is very long, you won't fit in so many cycles into one second. Conversely, if the wavelength is very short, you will get a large number of cycles per second. This means that the longer the wavelength is the lower the frequency will be, and the shorter the wavelength is the higher the frequency will be.

Mathematically, we say that the velocity (speed) of the radio wave is equal to the frequency multiplied by the wavelength. We write this as:

$$v = f\lambda$$

If you are good at maths you will know that you can rearrange equations so that if you know two of the three parts you can calculate the third, hence:

$$f = v/\lambda \quad \text{and} \quad \lambda = v/f$$

This is not a maths book, so let's make things a bit easier! If you can remember a simple triangle you can forget the equation and still get the answers. In the triangle shown in **Fig 72**, 'v' is always on top. It is also always the same number, 300. The 300 represents the velocity of the radio waves, which is 300 million metres per second, the same as the speed of light. Frequency (in MHz) and wavelength (in metres) are on the bottom and if you know one, you can work out the other.

Let's try a few examples:

- A transmitter is operating on a wavelength (λ) of 100m. What is its frequency?

  You don't know the frequency, so you cover 'f' on the triangle. What remains is 300 over λ, or 300 divided by λ. In this example, 300/100 = 3MHz. Using 300 for v always gives the answer in MHz.

- A receiver is picking up a frequency of 1.5MHz. What is the wavelength corresponding to this frequency?

  This time we cover λ, which leaves us with 300/f. In this case 300/1.5 = 200m. Remember, if you use 300 on the top and f is in MHz, the wavelength will be in metres.

- Here's a more tricky question. What is the wavelength of a signal at 10GHz?

  OK, cover λ, as we don't know the wavelength. That leaves 300/f, but in this example f is not in MHz, so you need to convert it. 10GHz is equal to 10,000MHz (remember the extra three zeros). The triangle now reads 300/10,000 = 0.03m or 3cm, a very short wavelength.

- Try calculating the wavelength of the 0.136MHz band, then imagine building a half-wave dipole! If you get stuck, ask your instructor or a more experienced amateur.

Radio amateurs often refer to the various bands allocated to them in terms of metres, because wavelength is an important concept. For example, you might hear talk of 40m, which refers to the 7MHz band. If you do the calculation, you discover that 40m is actually 7.5MHz (not 7MHz), so the name 40m band is not technically correct but it is nonetheless used throughout the world. The same applies to 145MHz. Is it really 2m?

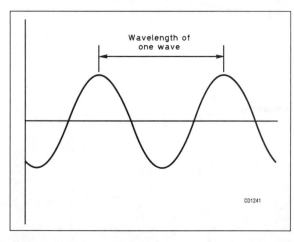

Fig 71: One wavelength is the distance from one peak to the next.

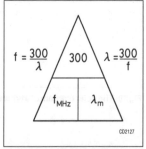

Fig 72: The formula triangle for calculating frequency from wavelength and wavelength from frequency.

Skip distance & skip zone
(6a.4)

AN HF RADIO SIGNAL can be received in two ways. It may travel to the receiver directly as a 'ground wave', or it can be reflected off the ionosphere, which we learnt about for the Foundation Licence. Ground wave is a signal that travels directly to the receiver without being reflected. However, the ground wave is gradually absorbed by the ground and the higher angle radiations travel off into space because the radio wave travels in straight lines and the earth is curved. Ground wave coverage is therefore fairly limited.

When a signal is being reflected from the ionosphere, the total distance from the transmitter to the shortest point where the signal is reflected to is called the 'skip distance' (i.e. the distance that a skip off the ionosphere will carry the signal). For HF signals, skip distances of up to 2000km are common. This means that the HF bands are generally used for nationwide and international contacts, rather than local, cross-town QSOs.

There is an area over which a signal will not be heard, because it is outside the range of ground wave but too close to be within the skip distance. This area is called the 'skip zone' or, in some textbooks, the 'dead zone'. All this is illustrated in **Fig 73**.

The ionosphere
(6a.1, 6a.2, 6a.3 & 6a.5)

AT THIS LEVEL we need to investigate the nature of the ionosphere in more detail. The ionosphere is composed of air, but it is much thinner than we are accustomed to on the earth's surface. Like other substances, air consists of molecules, which are the building blocks of matter. At the height of the ionosphere above the earth, the molecules are exposed to high levels of radiation from the sun. Because of the radiation that the molecules receive, they become 'ionised'.

The radiation that causes this ionisation is mostly of the type known as ultraviolet. You don't need to know the scientific details of ionisation, just that it causes a reaction that tends to split up the molecules of air. In doing so it creates ionised particles that reflect radio signals.

You will recall that the ionosphere ranges from 70km to 400km in height. In reality, the ionosphere is not just one layer in the atmosphere, there are several layers whose composition and height vary.

The layers are given letters, D, E and F, F being the highest (see **Fig 74**). The most important for HF communication is the F layer, which is responsible for the reflections that make most long-distance communication possible.

As you might expect, because ionisation is caused by solar radiation, there is quite a big difference between day and night. Solar radiation is at a maximum during the daytime. There is also a seasonal variation, because of the lower solar radiation received during winter compared with summer.

The other layer that has a marked effect on HF communication is the D layer. This is lower in the atmosphere, where the molecules are closer together. When they are ionised they tend to absorb radio waves, rather than reflect them. The effect is quite dependent on frequency, and typically anything below 6MHz will not break through. This means that LF propagation is limited to ground wave during the day.

However, the D layer disappears fairly quickly at night, when the sun's radiation is no longer present. This means that lower frequencies can then be reflected from the F layer and longer distances covered.

By contrast, the molecules in the F layer react rather more slowly to the onset of night. They are likely to remain ionised for several hours after sunset, continuing to provide reflections on both higher and lower frequencies.

The E layer appears to have little specific effect on HF propagation, but a form of propagation that takes place at VHF is called 'Sporadic-E'. As you might expect, it is associated with the E layer of the ionosphere. There are several theories about what causes Sporadic-E, but it is mostly a summertime phenomenon and may be linked to the weather.

It seems that large clouds of highly ionised gas form in the E layer and then move about rapidly. Contacts often have to be completed quickly, because the clouds will not consistently reflect a given signal path. The lower VHF region is most commonly reflected, usually up to around 150MHz. Because the E layer is lower than the F layer, signals will not be reflected quite so far, but multiple skips are possible.

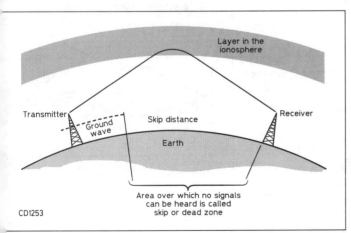

Fig 73: Illustrating skip distance.

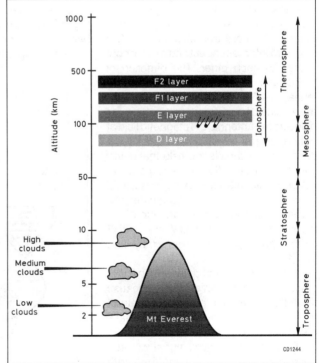

Fig 74: An overview of the ionosphere.

Sunspots (6a.2)

FOR HUNDREDS of years astronomers have recorded the number of sunspots on the sun's surface. The records show that there is cycle of sunspot activity, with maximum numbers occurring every eleven years. Over the last one hundred years a clear link has been established between the number of sunspots and radio propagation. In a nutshell, an increase in the number of sunspots indicates an increase in the solar activity that causes ionisation in the ionosphere. Increased ionisation normally improves HF propagation.

Polarisation (5f.4 & 5f.5)

YOU WILL recall that the polarisation of a radio wave is determined by the orientation of the antenna - a vertical antenna radiating vertically polarised radio waves and a horizontal antenna producing horizontal polarisation.

You now need some understanding of why that should be. Radio waves are one type of electro-magnetic radiation - light, infrared and ultraviolet being other types. Each of these types of radiation involves two fields; an electrical field (denoted by the letter 'E') and a magnetic field (denoted by the letter 'H') - hence the term 'electro-magnetic' radiation.

As your RF current flows through your antenna the two fields are generated around it to form a radio wave which propagates away from the antenna. The two fields are interdependent and exist at right angles to each other. The direction of propagation is at right angles to the antenna. See **Fig 75a**. To put it another way, the electrical field is produced in the same plane as the antenna, the magnetic field at right angles to it. You should quickly deduce that it is the electric field that determines the polarisation of the radio wave.

You would imagine that signals would be received best when a transmitter and receiver use the same polarisation, and this is certainly the case on VHF and UHF where propagation is generally line of sight.

On HF, the effect of polarisation is not generally as noticeable. This is because the ionosphere is in a constant state of movement, reflections tending to mask the effect of the polarisation used. In fact it's likely that a received signal will vary quite markedly in terms of signal strength and polarisation because of the constantly changing ionosphere.

VHF and UHF propagation (6a.5, 6a.6 & 6a.7)

AS WE GO HIGHER in frequency, signals are less likely to undergo ionospheric reflection. Several factors have an effect, including the angle at which the radio signals reach the ionosphere, but there comes a frequency where radio waves are not normally reflected and pass directly through. This point is usually taken as 30MHz, the boundary between HF and VHF. However, in practice you will sometimes hear VHF signals being reflected from the ionosphere. Equally, there will be times when signals in the 28MHz band will not be reflected.

On VHF and UHF, day-to-day radio contacts will be line-of-sight transmissions. The ground tends to cause some bending of waves, so ranges are generally a little further than you would expect, up to 100km being feasible with Intermediate power levels. VHF and UHF contacts tend to be more local than HF.

On occasion, the troposphere can increase this range considerably. The troposphere is the layer of the atmosphere which is nearest the earth. Sometimes the troposphere can cause another type of bending of radio signals, under conditions known as a 'temperature inversion'.

Under normal circumstances, the temperature of the air in the troposphere decreases with height. We all know that mountains become colder as we go higher. Sometimes an area of warmer air becomes trapped higher in the troposphere, along with some humidity. When this happens signals may become trapped in a layer (or duct) of the troposphere and return to earth at a distance. This phenomenon known as 'tropospheric ducting'. See **Fig 75**.

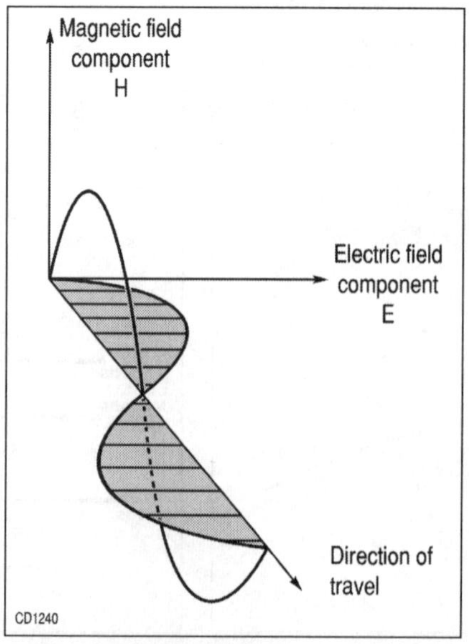

Tropospheric ducting can extend the range of VHF and UHF signals to several hundred kilometres. Because they are associated with patches of warmer air, temperature inversions occur more frequently in the summer.

You also need to be aware that the weather can also reduce VHF and UHF propagation. Clouds of rain and hail contain lots of water and/or ice that absorb shorter wavelengths. Severe weather conditions can prevent even local contacts on these bands.

Summary of typical distances

- LF & MF. Ground wave during daytime, can give national coverage (hundreds of km). May be reflected at night to extend range (thousands of km).
- HF: Ionospheric reflection gives fairly reliable international communications (thousands of km), best during the day.
- VHF & UHF: Line of sight, local contacts (up to 100km), can be enhanced by Sporadic-E (thousands of km) and tropospheric ducting (hundreds of km), but can equally be hindered by rain and hail clouds.

Further information

PROPAGATION is a fascinating subject and one of the best publications to read on the subject is the RSGB book *Radio Propagation - Principles and Practice*.

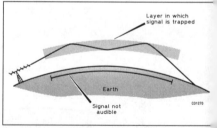

Fig 75: How a tropospheric duct can carry VHF signal

Fig 75a: E and H fields are at right angles to each other and the direction of travel

RF Coax Connectors

Fitting an RF plug (10d.7)

DURING YOUR Foundation training you met a couple of RF connectors, the PL259 and the BNC. You may already have used them as part of your M3 station, but you may not have fitted them yourself. The object of this exercise is to show you how to fit the plugs and so prepare you for one of the practical assessments.

You must be able to demonstrate to the assessor that you can prepare coax and solder an RF plug on it. The type of plug is not specified, but the two mentioned above are the most common. If your equipment uses 'N' type connectors it would be sensible for you to use one of those. Please note that 'phono' and Belling-Lee plugs are not acceptable for the assessment.

Details of fitting various RF plugs can be found in many of the books available on antennas, but details of the PL259 and BNC are set out here.

PL259

THERE ARE SEVERAL types of PL259. The instructions below are for a plug that includes a reducer to fit the size of coax being used and should give you the basic idea of the procedure, even if your plug is slightly different.

1. Prepare the reducer by cleaning the plain end just beyond the thread using wire wool, a sharp knife or a small file.

2. Holding the reducer in a vice, a wooden clothes peg or a pair of pliers with a rubber band around the handles, tin the area you have just cleaned. You will need to use a fairly large bit and a 25W soldering iron. Beware! The reducer will remain very hot for some time after you take the iron away, so leave it to cool for a couple of minutes.

3. Thread the reducer over the coax

4. Prepare the end of the coax by removing about 30-40mm of the outer covering, using the same technique as for mains cable. Be very careful not to cut into the fine strands of the braid, as they can break off and cause short circuits inside the plug.

5. Using a small screwdriver, tease out the strands of the braid right back to the outer covering.

6. Trim the braid back to about 25mm from the end of the outer covering using a pair of sharp scissors, then fold them back over the reducer. Try to spread them evenly around the reducer (see **Fig 76**).

7. Hold the coax in a vice so that you can work on the reducer.

8. Tie a few turns of cotton around the braid to hold it in place and then solder it to the reducer. Only use the minimum amount of solder, as too much will prevent the reducer fitting inside the plug afterwards. After soldering, leave it to cool.

9. Thread the ferrule over the reducer so that the inner thread is towards the prepared end of the coax.

10. Cut the inner insulation about 3mm away from the braid/reducer and pull it off. Be very careful not to cut into the inner conductor. If the inner is damaged you will need to crop the coax and start all over again.

11. If the inner conductor is stranded (some coaxes have a single, heavier gauge wire), twist them together and tin the end to prevent any of the strands bending back when you get to the next step.

12. Guide the inner through the body of the plug until it appears out of the end of the pin, checking that all the strands have come through.

13. Screw the body of the plug onto the reducer and tighten it with a pair of pliers.

14. Solder the inner conductor to the pin. Make sure that you do not get too much solder on the inside, as it can flow up the pin and short the pin to the braid; and too much on the outside of the pin, or it will prevent it from fitting in a socket.

15. Trim off any excess inner conductor and screw the ferrule into place.

16. Using a multi-meter set to a high resistance range, check that there is no short circuit between the body of the plug and its pin. If all is well you can use the coax and plug. If not, chop it off and start again.

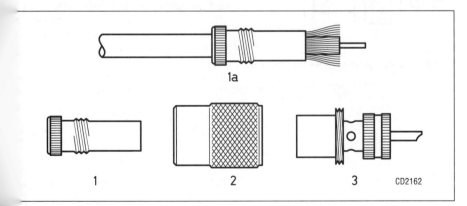

Fig 76: Exploded view of a PL259 connector.

BNC

AS WITH THE PL259, there are a number of variations on the basic design. These instructions cover one type. You may have to make some changes, depending on your actual plug.

1. Thread the nut, the metal washer and the rubber washer over the coax.

2. Prepare the end of the coax by removing about 30mm of the outer covering, using the same technique as for the mains cable. Be very careful not to cut into the fine strands of the braid, as they can break off and cause short circuits inside the plug.

3. Thread the cone over the braid up to the outer covering.

4. Using a small screwdriver, tease out the strands of the braid right back to the cone.

5. Trim the braid back to about 10mm from the cone using a pair of sharp scissors and then fold them back over the cone. Try to spread them evenly around the cone (see **Fig 77**).

6. Trim the braid so it just covers the cone.

7. Cut the inner insulation *exactly* 5mm away from the braid/cone and pull it off. Be very careful not to cut into the inner conductor. If the inner is damaged you will need to crop the coax and start all over again.

8. If the inner conductor is stranded (some have a single, heavier gauge wire), twist them together.

9. Cut the inner conductor *exactly* 4mm from the inner insulation.

10. Holding the coax in a vice, a wooden clothes peg or a pair of pliers with a rubber band around the handles, fit the pin over the inner conductor and solder it in place. This is best done by heating the pin and feeding the solder through the hole in the side of the pin. Allow the pin to cool for a minute or so.

11. Place the body of the plug over the pin, being careful not to disturb the braid on the cone. Satisfy yourself that the pin has seated correctly in the insulation and is protruding from the front of the plug.

12. Gently push the rubber washer up against the cone and then the metal washer against the rubber and screw the nut into the body of the plug.

13. Tighten the nut with a pair of pliers.

14. Using a multi-meter set to a high resistance range, check that there is no short circuit between the body of the plug and its pin. If all is well you can use the coax and plug. If not, chop it off and start again.

Further information

DETAILS OF FITTING various RF plugs can be found in the book *Backyard Antennas*. The book and suitable RF connectors are available from the RSGB shop.

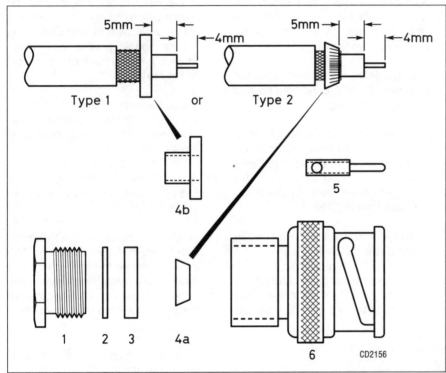

Fig 77: Exploded view of a BNC plug. Note that differences exist between plugs from various manufacturers.

Electromagnetic Compatibility (EMC)

IN THE FOUNDATION course you learned that EMC was the avoidance of interference between two items of electronic equipment and that radio transmitters could be a source of interference. You learned that good practice starts in the radio shack and looked at some of the ways that a transmitted signal could be conducted or radiated to a device that was suffering from interference. You are advised to look over that material again, as it will help you understand the topics introduced here.

The EMC Directive
(7a.1, 7a.2 & 7a.3)

THE BASIC PRINCIPLES of achieving electromagnetic compatibility are to make electronic devices so that they do not radiate too much unwanted RF energy and for all electronic equipment to be made immune to picking-up unwanted signals. A piece of European legislation called the 'EMC Directive' makes this a legal requirement and, in general, it works very well. Lots of complex electronic equipment does indeed work side by side with other equipment without problems.

So why is EMC an issue? Our hobby and all other radio users deliberately produce lots of RF energy from our transmitters, so there is a possibility of some incompatibility.

Most professional radio users have transmitter sites well away from domestic buildings and other sensitive areas such as the medical electronics in hospitals, so there are few problems. However, most amateurs need to put their transmitters in domestic areas where there are lots of devices designed to be very sensitive to radio signals; broadcast receivers, televisions, baby alarms, cordless telephones, etc. If you place any of these devices near to powerful RF transmitters they *will* pick up signals they are not designed for. The EMC rules do not really allow for these situations - there is a limit to immunity!

In the domestic environment it may be necessary to provide extra protection (immunity) for affected devices, or it could turn out that the transmitted power will

need to be adjusted down to below the limit actually permitted by the Intermediate licence. That said, new electronic equipment that complies with the EMC directive is much more immune to interference than older equipment and, when properly installed, does not normally give rise to problems. If new equipment does suffer, it may be that the installation is poor. However, that is something the owner of the equipment cannot really be expected to understand, especially if his equipment works until you transmit.

You may have come across a number of different types of television downleads and amateur feeders. The cheap ones have hardly any copper braid to prevent the signals leaking out, *or in*. A poor installation using cheap cable can easily ruin the performance of even the best equipment, especially if it is not waterproofed and the rain corrodes what little copper there is. That also applies to the quality of installation in your shack.

Recognising interference
(7c.1 & 7c.2)

YOUR OWN TV and radio/music system is closer to your transmitter than those of the neighbours and you should be particularly critical to ensure their performance is faultless and interference free. So what are we looking for?

Amateur radio interference to televisions depends on the type of transmission and the affected device. Similar effects are also possible from taxis and delivery vans fitted with two-way radios and nearby commercial or emergency services transmitters.

- FM (VHF or UHF): Wavy, herringbone patterning on TV, possible loss of colour. Normally no effect on sound, but severe cases may cause distortion or loss of sound.
- AM or SSB: Similar patterning on screen, possibly in time with speech. Distorted voice-like sounds, can be intelligible.
- Morse code: Possibility of flickering picture or wavy effects, possibility of clicks or distortion on sound.

Amateur radio interference to music systems and telephones is less common, but can cause problems where it does happen.

- FM: Very low possibility of distortion on music systems and telephones.
- AM or SSB: Distorted voice like sounds on music systems and telephones, can be intelligible.
- Morse code: Low possibility of clicks or distortion on sound.

Interference can come from other sources, and radio amateurs can suffer from EMC problems just as much as they can cause them!

- Electric motors: Spots and lines on TV screen. Possibility of buzz on the sound.
- Thermostats: Bursts lasting 2-10 seconds of spots and lines on screen, possibly quite intense, and a rasping noise on sound like screwing up paper. Bursts repeat at intervals of a few minutes, each time the thermostat switches.
- Vehicle ignition: Spots/lines on screen. Clicks on sound in time with engine speed.

Digital televisions are affected in a very

different way. Good immunity can be expected if the TV signal is strong, but problems may occur in weak signal areas. Since the picture is encoded and digitally compressed in a manner not unlike zipping or compressing a computer file, the effect is much less predictable and the effects are similar to the wanted signal being too weak.

In mild cases the picture may become jerky or occasionally break up. It may also appear as if it is made up of square tile-like blocks. Those with an analogue TV may have seen similar effects when watching outside broadcasts, particularly from mobile cameras such as those fitted to racing cars to give a drivers' view. The sound may also fail, but this is much less likely.

Interference mechanisms

THERE ARE SEVERAL ways that interference can find its way into electrical and electronic devices. You need to know a bit more than you covered in your Foundation training, but this is a big subject and you will find out more when you move up to the Full licence level.

Direct pickup (7c.3)

THE RF TRANSMISSION is picked up inside the affected device and affects the sound or picture. Usually the exact frequency of either the transmission or the frequency that the device is tuned to (if it is a radio or TV) does *not* affect the severity of the interference.

Direct pickup is difficult to cure and usually requires modifications inside the affected equipment. It should therefore be left to skilled service personnel. Neighbours' equipment must *never* be modified, since any subsequent fault will be blamed on you and could have much more serious consequences.

More commonly it is the various cables leading to the affected device that act as aerials, carrying the unwanted RF into the equipment and causing the problem. A ferrite ring filter may be fitted on any leads, audio or power, to keep unwanted RF out

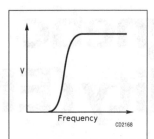

Fig 78: The characteristic of a high pass filter.

of the affected equipment. Adding external ferrite rings should not affect warranty.

Television amplifiers (7c.4)

IN WEAK SIGNAL AREAS amplifiers are often used to boost the television signal. These are usually bad news for the amateur.

Weak TV signals mean that your transmissions will appear much stronger by comparison. To make matters worse, many of these amplifiers have a very wide bandwidth, far wider than the range of frequencies used by television transmitters. They often cover amateur bands, particularly the 2m and 70cm bands, so your strong amateur signals are also amplified, overloading either the amplifier or the television and causing severe interference, possibly with total loss of the picture or sound.

The cure for this depends on circumstances. The first check is to see if the amplifier is needed at all, so try removing it. Most have a male socket on one end and a female one on the other, so it is simple to remove it and plug the downlead straight into the television. If it is a neigh-

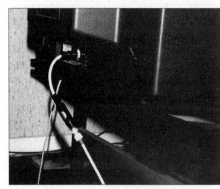

A high pass filter fitted to the back of a TV.

bour's TV, ask them to remove it or at least work with the neighbour's co-operation. Now try transmitting, pointing rotatable antenna in several directions. If you are lucky, that may be the end of the matter. If a booster is found to be necessary you should check to see if the one being used is correct for the TV transmissions in your area. Simply replacing a broadband amplifier with one for the specific band in use can remove the problem. If not, and it is VHF/UHF frequencies that are a problem, then a high-pass filter (**Fig 78**) may be effective. This will pass TV frequencies (470-845MHz) and everything above, but reject the lower frequencies of the amateur UHF (430-440MHz) and VHF (144-146MHz) bands.

The filter should be plugged in at the back of the TV and the downlead plugged into the filter input, as shown in the photo below. If the TV signal amplifier/booster is fitted to the back of the TV and is required, the filter should be placed before the amplifier by removing the downlead from the amplifier, inserting the filter at the amplifier input and connecting the downlead to the filter as before. The link between the amplifier and TV is left untouched. In this way the offending signals into the amplifier are minimised.

Downlead pickup (7c.5)

IF HF TRANSMISSIONS are a problem, the TV aerial amplifier is less likely to be the culprit. It is more likely that the HF transmission is being picked up by the braid of the TV downlead and conducted into the TV along it.

A ferrite ring will be of use here, but take care not to kink the downlead. A larger diameter ring may be needed to avoid kinking and to get sufficient turns on it. It may help to use two rings stacked and wound as if there was a single ring. The TV signal will be unaffected by a ferrite ring on the downlead.

For HF and VHF protection, both filters might be needed. Remember – there are so many different scenarios that it is impossible to say in advance just what the effect (or cure) might be!

Other non-amateur problems (7c.5)

IT MAY BE THAT a complaint by a neighbour, or a problem to your own equipment, say your amateur receiver, is traced to a non-amateur source, perhaps an electric drill or a faulty thermostat.

Having found the offending item, what can be done? Well, that is really up to the owner, but the general procedures would be as follows:

- Thermostats should be replaced, but since they are connected to the mains, unskilled repair should not be attempted, as it is simply not safe.
- Electric drills and other motors such as lawnmowers or sewing machines should be properly filtered. A new device should be returned to the supplier. Older devices may have just developed a fault and the safety implications of continued use should be carefully considered.
- Other devices may have always been a bit 'noisy' but not caused a problem until a new TV or a more sensitive radio was purchased. A proprietary mains RF filter may cure the problem.

More EMC detective work (7c.6)

SUPPOSE YOU HAVE identified that your transmissions are causing a problem and want to know a bit more about just how the signal is getting into the affected equipment.

One good check is to replace the station antenna with a dummy load and transmit on the frequency that has been causing trouble. If the problem ceases then it must have been caused by the strength of the radiated signal.

Maybe you can move the antenna further away, maybe a ferrite ring on the affected device would be sufficient, maybe you should consider not pointing your beam at the neighbours house, or you may have to think about reducing your transmit power.

However, if the problem persists with the dummy load in place it was not your radiated signal that was causing the problem. If the transmitter has a 12V power supply, try a test using a battery instead. A car battery will do, but don't use long leads to the car as they may radiate. Put the battery where the mains power supply was sited and try that. If this cures the problem, the RF was being conducted along the mains and more filtering or the screening of power leads is called for.

Further information

PROBABLY THE BEST reference book on this topic is the RSGB's *Guide to EMC* by Robin Page-Jones, G3JWI. It contains a good mix of technical theory and some practical solutions to common problems.

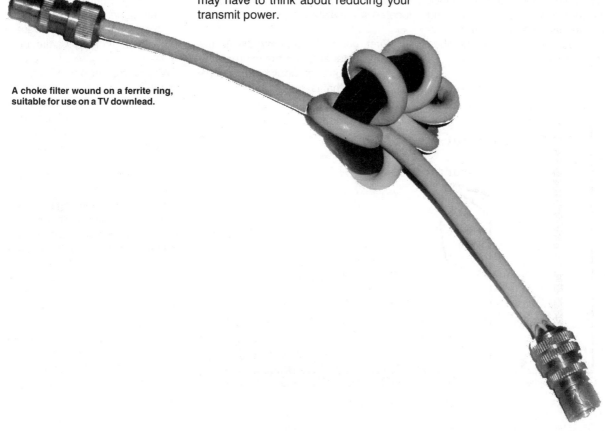

A choke filter wound on a ferrite ring, suitable for use on a TV downlead.

Checking for Harmonics & Spurious Emissions

Why check? (2f.3)

ONE OF THE licence conditions requires you to check your station for 'unwanted emissions' from time to time. Unwanted emissions may be harmonics (see the transmitters worksheet) or spurious emissions from a faulty transmitter. You should also check to be sure you are not causing interference to your neighbours' radio equipment or to any other users of 'wireless telegraphy'.

A well designed and built transmitter should not produce any harmonic radiation and the low pass filter in the output should minimise what little is present. However, you must still check your transmissions. There is always a small chance that your transmitter may be producing spurious signals unrelated to the transmission or perhaps only present when the carrier is modulated.

How to check

A PIECE OF TEST equipment that was popular many years ago is the wavemeter. Basically this is a crystal diode receiver that can be tuned over the transmitted frequencies plus the second and third harmonics. Today the greater need to avoid interference to others and the advances in transmitter technology to minimise spurious and harmonic signals means that the wavemeter is not sensitive enough to be of any practical use.

The simplest method is to use a receiver that covers the frequencies of interest. If you are using a transceiver you may need to borrow a separate receiver. You may also wish to recruit another amateur located some distance away to help you.

1. First, connect a dummy load to the transmitter and turn the microphone gain to zero or unplug the microphone if that is easy to do.

2. The receiver should be placed a short distance away, set to receive Morse (CW) and tuned to the same frequency as the transmitter, but without any antenna connected.

3. Switch the transmitter to CW and transmit whilst checking the receiver. A strong signal should be received.

4. Connect a well-screened dummy load to the receiver's antenna connector (a low power one will do fine) and check to see that the received signal level remains less than full scale on the receiver's signal strength meter. Alternatively, move the receiver further away and then see if the signal drops. This is one of the tests where you might want to check with another local amateur.

Why is this important? If the received signal is too strong, the receiver will be overloaded. An overloaded receiver circuit can also produce harmonics inside the receiver. If that happens you will not know if the harmonics you are detecting are produced in the transmitter or the receiver. However, if your receiver is indicating less than S9 on the meter then it is unlikely to be producing its own harmonics. Consequently, any harmonics or spurious signals you do detect can be blamed fairly and squarely on the transmitter.

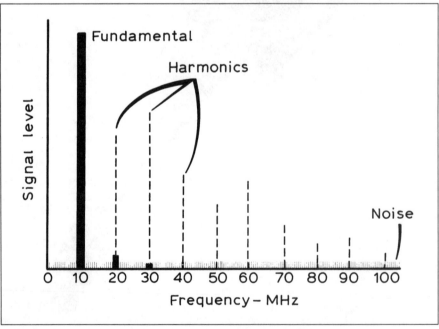

Fig 79: Frequency spectrum showing a fundamental at 10Mhz and its harmonics.

Checking for harmonics (4e.2 & 4e.3)

You will recall that harmonics can cause interference to other amateur bands (e.g. the second harmonic of a signal on 7.050MHz will be on 14.100MHz where amateur beacons transmit) or to other radio users (e.g. the second harmonic of 50.50MHz will be on 101MHz, a popular VHF broadcast frequency). You can check for harmonics as follows:

1. Calculate the frequencies of the harmonics, that is 2, 3, 4 and 5 times the transmitted frequency, and make a note of them. **Fig 79** gives an example.

2. Tune the receiver to each of those frequencies and listen for any signals from your transmissions. It is quite possible that you will hear something, but it should be very weak.

If it is strong, move the receiver further away until the received level of the *transmitted* frequency is definitely below full scale on the receiver's signal strength meter and try again. If the harmonics are still almost as strong as the fundamental it suggests a possible problem with the transmitter, in which case it is time to get help from a more experienced amateur or speak to your equipment supplier.

Checking for overmodulation (4e.1)

IF YOU CAN, try some tests with the other modes such as SSB and AM. If they are fine, the odds are that FM will be fine, but you can try that too. Don't forget to turn the microphone gain back up, or these tests will be wasted.

Whilst on SSB, quite a critical test is to try off-tuning the receiver by up to 12kHz either side of your transmissions. However you really must be sure that the receiver is not being overloaded for this test. It must be comfortably off full scale on the received signal strength indicator. The help of another local amateur would be useful for these tests.

Ideally, once you are more than 3 or 4kHz away from the transmission, nothing should be heard at all. If you can hear weak voice-like signals, try reducing the transmit power slightly and then the microphone gain. If you find that one of those adjustments causes the received signals to come and go quite suddenly, you have found the point at which the transmitter is overloading or where the modulator is overdriving. Make a note not to set the transmit power or microphone gain too high. If you feel unsure, your instructor or another amateur may be able to help.

Interference from CW (4e.5)

Not all transmitter interference comes from voice transmissions. Problems can also arise from poor CW keying.

As you know, Morse code is sent by switching a continuous RF signal on and off to form the various characters. If the rise of the RF on key down, or the fall on key up, is too rapid a square shaped RF envelope will be produced. Such a waveform is likely to generate excessive bandwidth similar to the overmodulation of an am transmission.

In a good CW transmitter a filter will be built into the keying stage. This filter will slow the rise and fall of the RF envelope to maintain the narrow bandwidth of a good CW transmission.

Checking for spurious emissions

FINDING SPURIOUS SIGNALS can be quite a game - they could be anywhere. The solution to this is to use a receiver to tune around the bands. The trouble is, there are likely to be strong broadcast signals that are picked up despite not having a proper aerial connected.

The test is a lot easier if a friend can key the transmitter, as if sending Morse. That way, if the signal is related to your transmission it will be obvious since the received signal will come and go in time with the keying of the transmitter.

After the tests

MAKE CAREFUL NOTES of these proceedings. Apart from the fact that you should log these tests to comply with the licence conditions, they will be of real help if you are asking a more skilled amateur, a retailer or the RA for assistance. They are likely to be impressed and realise that you are taking care of your equipment.

Ideally these tests should be repeated on each band that you use. Once mastered it does not take too long and can be carried out over a period of time, logging each set of tests as they occur.

If you are a member of a radio club or know a friendly retailer with test and repair facilities, they may be able to check your transmitter much faster. They may have a piece of test equipment that tunes rapidly across the whole band, or even many bands, and displays the results on a screen. However, such test equipment (which is known as a 'spectrum analyser') is far too expensive for the average amateur.

Good Radio Housekeeping

THE CONCEPT BEHIND good radio housekeeping is to have a well designed, well laid out, well assembled shack so as to minimise the RF energy that finds its way into other items of electronic equipment. Such equipment could include the neighbours' TV, video, radio, telephone, computer, sound system and all the other gizmos that seem to fill the house.

Layout and cables (7b.1 & 7b.2)

A GOOD LAYOUT keeps RF and power or audio/microphone leads as far apart as possible. RF leads should use good quality coaxial cable and be fitted with good quality plugs and sockets. Microphone and other audio cables must also be screened and again both the conductors and the screen should be properly joined so as to maintain the continuity of the outer screen.

To prevent any RF from your transmitter entering the mains supply, the mains leads to the power supply unit and the low voltage DC power leads should be fitted with ferrite rings as close to the equipment as possible (see **Fig 80**).

Even a good transmitter may have some spurious signals, that is low level outputs on frequencies unconnected with the intended transmission. Alternatively, it may produce harmonic radiation. Fitting an external low pass filter that will allow the wanted signal to pass but reduce the higher frequency harmonics can help. If required, this should be inserted after the SWR meter and before the ATU.

Antennas (7b.5)

IN YOUR FOUNDATION training you learned to keep the antenna as far away from houses as possible. This advice is even more important if you are intending to run the higher powers allowed by the Intermediate Licence.

In towns and cities the temptation to put your transmitting antenna in the loft should be resisted. Often there is a lot of house wiring in the loft, the main TV reception antenna may even be there, and the potential for RF being picked up should not be under estimated. Similarly, a temporary installation with an antenna in the shack is almost guaranteed to cause problems.

It is worth repeating that antennas should be kept well away from the house and balanced antennas are less likely to cause interference. You should also be aware that vertical antennas have a greater potential to cause interference and need a very good earth at their base, possibly with a number of earthed radial wires buried under the lawn.

Earthing requirements (7b.3)

THERE ARE TWO different earths in the amateur shack, each with a different purpose. Confusing these matters may cause other equipment to suffer interference from your transmitter, but far more importantly it may create a serious risk of electric shock, and that could be fatal!

The first type of earth is the mains earth. All mains powered equipment must be earthed for safety reasons. The only exception to this rule is where equipment is 'double insulated', which will be shown by the symbol in **Fig 81**.

Since most amateur radio equipment has some exposed metalwork it is very seldom double insulated, so it must be earthed using the earth pin in the 3-pin mains plug. This earth is for *your* safety, so don't remove it!

The second type of earth is the RF earth, which is there to divert RF currents away from the mains earth and safely down to ground. It consists of one or more earth rods driven into the ground, preferably in damp soil. These rods are made for the purpose and are available from both amateur antenna/feeder suppliers and good electrical suppliers dealing with the trade and local electricians

These rods are some two metres long and are usually made from copper coated steel. They should be hammered into the ground close to the shack, taking care to avoid hidden services and drainage pipes. A heavy cable should be clamped to the rods and connected directly to the transmitter or Antenna Tuning Unit, as shown in **Fig 82**.

Log keeping (7d.1 & 7d.2)

ALTHOUGH LOG KEEPING is a requirement of your licence, that is not really why you must keep a log. The licence requirement is based on good radio housekeeping.

Amateurs are the only radio users allowed to design and build their ow

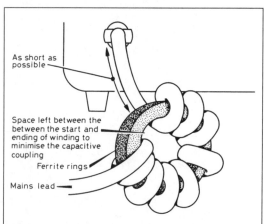

Fig 80: A ferrite ring choke on a mains lead.

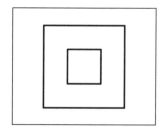

Fig 81: The double insulated symbol on some mains powered devices.

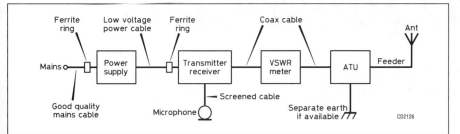

Fig 82: Station layout, showing filters on the power leads and the correct earth connection.

equipment. All other users must buy properly tested and approved transmitters. Amateurs are also allowed to transmit high powers in residential areas, with all the possibilities for interference. That really is quite a privilege and it is why training and exams are necessary.

The Logbook is very useful if an interference problem arises. Were you transmitting at the time of a problem? If not, you can quite quickly show you were not the source. If you were, what exactly were you doing? Such information will help greatly in recreating the same situation, when investigating what the cause of the problem is.

If the person making the complaint is also willing to keep a log, it will help in identifying what transmissions are causing difficulties. If the matter becomes a formal complaint to the authorities, i.e. the Ofcom local office, they will ask for logs to be kept as a first action in identifying the cause.

In such cases an honest, open and professional approach is the best policy. Ask the person making the complaint to co-operate with the tests to identify the problem. Be as tactful as you can. You need to understand that they do not want the interference and they need to understand that you want to be able to transmit and enjoy your hobby. Show you are serious about resolving a case of interference, but also serious about being able to pursue your hobby.

By all means offer not to transmit at key times until a cure is found, but do not admit blame and do not make the offer sound like a permanent change. If the problem is caused by inadequate immunity in neighbours' equipment, as it often is, joint tests may demonstrate that without the difficulty of having to say as much. It is always useful to be able to show that your own TV and radio equipment is free from interference.

Seeking help with problem cases (7d.3)

IF YOU TRY everything but are still having problems, you may need assistance. The best source is the RSGB's EMC committee. The EMC committee produce several leaflets on EMC and interference. Some are aimed at the amateur, whilst others are intended for someone making a complaint. They all aim to provide unbiased and accurate information on interference matters. The tips in those leaflets, which are available by post or can be downloaded from the EMC Committee website, may help you solve a problem. If not, RSGB members may contact the EMC committee chairman who will provide the address of the nearest committee member who can assist.

Ofcom also publish advice leaflets and Ofcom's local office staff can investigate problems to licensed radio services such as television reception. This should be regarded as a last resort. The person making a complaint should register their problem by completing the form in the appropriate Ofcom information leaflet and sending it to the local office (address in the leaflet). It is a chargeable service, but if the installation of the person making a complaint is in good order and a problem turns out to be outside that person's control or it is found to be coming from an unauthorised source, the charge may be waived. Full details are in the leaflets.

Revision Questions 2

BY NOW YOU should be familiar with the exam question format. Here are some more questions from the second half of the book to get you thinking about the material.

1. **In a single sideband transmitter, the modulator is known as:**
   (a) The modulating transformer
   (b) A varicap diode
   (c) A balanced modulator
   (d) The keying stage

2. **If the PA transistor in a CW transmitter is drawing 100mA from a 100volt DC supply, what is its input power?**
   (a) 200W
   (b) 10W
   (c) 200mW
   (d) 10mW

3. **Which mode would allow you to operate with your frequency dial closest to a band edge?**
   (a) CW
   (b) FM
   (c) SSTV
   (d) FSTV

4. **Which of the following is NOT considered to be 'Unattended Operation'?**
   (a) Operating a beacon
   (b) Leaving a packet station running
   (c) Controlling a transmitter through an RF remote control
   (d) Controlling a transmitter by a hard wired computer

5. **Which of the following is required to receive SSB?**
   (a) An RF amplifier
   (b) A frequency discriminator
   (c) An audio filter
   (d) A carrier insertion oscillator

6. **Balanced feeder cannot be run along walls because:**
   (a) It has lower losses than coax
   (b) It is not insulated
   (c) The wall could cause the RF currents to become unbalanced
   (d) The RF could leak through the wall and cause interference

7. **Which of the following modes does the Intermediate licence NOT permit on the 14MHz band?**
   (a) SSTV
   (b) FSTV
   (c) PSK31
   (d) RTTY

8. **Which part of a mains power supply unit changes AC to DC?**
   (a) The transformer
   (b) The diode
   (c) The capacitor
   (d) The fuse

9. **If you feed 50W of RF into a Yagi with a gain of 6dB, what would the ERP be?**
   (a) 50W
   (b) 100W
   (c) 200W
   (d) 300W

10. **An antenna tuning unit normally:**
    (a) tunes the antenna to frequency
    (b) changes the antenna feedpoint impedance
    (c) changes the impedance at the transmitter
    (d) reduces the transmitter power to a safe level

11. **The layer in the ionosphere that is mainly used for HF DX is the:**
    (a) D-layer
    (b) E-layer
    (c) F-layer
    (d) Ozone layer

12. **Which of the following is the MOST likely cause of television interference from an HF station?**
    (a) Direct pick up by the TV antenna
    (b) Direct pick up through the TV case
    (c) RF pickup on the TV antenna downlead
    (d) Mains borne interference

13. **Which transmitter design idea would reduce the chance of radiating harmonics the MOST?**
    (a) Using a crystal oscillator
    (b) Using a balanced modulator
    (c) Keying the power amplifier
    (d) Fitting a low pass filter

14. **Who is permitted to receive QSL cards through the RSGB bureau?**
    (a) Any licensed radio amateur
    (b) Licensed radio amateurs who are members of the RSGB
    (c) Anyone
    (d) Anyone who is a member of the RSGB

15. **Which of the following would include an IF amplifier?**
    (a) A crystal diode receiver
    (b) A tuned radio frequency receiver
    (c) A superhetrodyne receiver
    (d) A super regenerative receiver

16. **If a coax is said to have a 10dB loss, how much of a 50W transmission would reach the antenna?**
    (a) 500W
    (b) 50W
    (c) 25W
    (d) 5W

17. **If a mains transformer has more turns on the primary winding than the secondary winding, it will:**
    (a) change AC to DC
    (b) change DC to AC
    (c) step up the voltage
    (d) step down the voltage

18. **Referring to a Yagi antenna, which of the following is true?**
    (a) The radiation pattern is omnidirectional
    (b) All the elements are the same length
    (c) The reflector is longer than the driven element
    (d) The director is longer than the driven element

19. **Which of the following has the greatest effect on UHF propagation?**
    (a) The sunspot cycle
    (b) The time of day
    (c) The time of year
    (d) The weather

20. **Which of the following would be the MOST effective cure for mains borne interference to a television from a VHF transmitter?**
    (a) Moving the antenna further away from the affected equipment
    (b) Fitting a low pass filter at the transmitter
    (c) Fitting a high pass filter in the TV downlead
    (d) Winding the transmitter power cable around a ferrite ring

How did you get on?

IF YOU ANSWERED all these questions without difficulty you are probably ready for the Intermediate exam. If some cause you to go back and read the appropriate worksheets again, you may have some revision to do. In either case, revision time will not be wasted! The more you read through the worksheets, the more you will understand and remember for the exam.

All that remains are the practical and written assessments. If you are well prepared, you have nothing to fear.

The Practical Assessment

IN COMMON WITH the Foundation Licence training, you *must* complete a practical assessment *before* you can sit the written Intermediate examination. How you are assessed will depend on how you studied.

If you attended a tutor-led course and have completed all the practical exercises in this book you have probably already been assessed on most of the practical work. On the other hand, if you have been studying alone, or if your instructor is not a registered assessor, you will need to attend a practical assessment *before* your written exam. In such cases it is expected that they will take place on the same day at the same location, one after the other.

What's involved?
(10d.1-8, 10e.1 & 10f.1)

THERE ARE SEVERAL syllabus objectives that can only be assessed through practical exercise. These are:

- To construct a simple DC circuit (Worksheet 5)
- To measure potential difference and current in a simple DC circuit (Worksheets 9 & 10)
- To demonstrate that a diode will only conduct in one direction (Worksheet 20)
- To demonstrate that a transistor can be used as a switch (Worksheet 23)
- To fit an RF connector to a piece of coax (Worksheet 32)
- To fit a 3-pin mains plug to a piece of 3-core mains cable (Worksheet 7)
- To measure the resistance in a number of resistors and check the colour code (Worksheet 14)
- To calibrate a variable frequency oscillator (Worksheet 18)
- To construct a radio related project (Worksheet 6)

How can I prepare?

THE ASSESSMENT IS intended to be a friendly affair, similar to the Foundation assessment. It should not be seen as a harsh test, more like an integral part of the learning.

Much of the construction work can be completed prior to the assessment, indeed the radio-related project must be completed beforehand. However, you will be questioned about the construction, so that the assessor can assure him/herself that it is all your own work. Be prepared to talk them through what you did, how you did it and any difficulties you experienced.

The assessor will want to see that you can carry out the measurements and the calibration of the VFO. If your project includes a VFO, you can use that for the assessment. If not, the assessor will provide a VFO for you to use.

If you haven't been able to complete the practical worksheets during your training

you should at least read them through before the assessment to have a good idea of what is involved. However, the assessor will explain everything before asking you to demonstrate your skills.

How long will it take?

THE TIME FOR THE assessment will depend on many things, not least how much preparation you have done. An assessment for a well-prepared and competent candidate may only take 30 minutes. At the other end of the scale, if all the assessment tasks have to be carried out on the day, it could take two or three hours. You may find that the assessment centre has organised practical assessments for a morning session with the written exam in the afternoon.

What about candidates who cannot complete the practical work?

THE RADIOCOMMUNICATIONS Agency recognises that some candidates will not be physically capable of completing some (or all) of the practical assessments. Assessors have been briefed to take disabilities into account. In such cases the assessor will carry out the assessment by asking the candidate questions about the practical activities. For example, a blind candidate would not be expected to fit a 3-pin mains plug, but they would be expected to describe stripping-off the insulation and where to connect each of the coloured wires so that an assistant could complete the task on their behalf.

The Written Examination

Is it like the Foundation exam?

THE INTERMEDIATE exam is like the Foundation in that it is a multiple-choice assessment. The structure of the questions is exactly the same as the Foundation exam; there is a statement, question or diagram, together with four possible answers. What you have to do is pick the correct one.

There are two differences between the exams. The first one should be obvious; you will be asked questions on the Intermediate syllabus. This will require you to have more knowledge than for the previous exam. If you have worked through this book you should have covered all that you need to know. The second difference at this level is that the exam has more questions, 45 to be precise. However, you will be glad to know that you are allowed more time to complete the exam.

What will the questions be about?

THERE ARE TEN TOPICS in the Intermediate syllabus, but some will have more questions in the exam than others (see **Table 6**). You will see that the first two topics have been joined together for examination purposes. However, all areas will be tested, so be sure that you have covered all the material.

How can I prepare for the exam?

THE BEST ADVICE is to go back through the worksheets in this book and make revision notes for yourself as you go. Don't be afraid to write in the margins if you find a way of remembering a key point, it's your book! Hopefully you will already have worked through the practical exercises. These were included to help make the theory real and to aid your understanding of some quite complex ideas.

How should I tackle the exam?

WHEN IT COMES TO the exam itself, the key is to read each question carefully. It is quite easy to see a question and jump ahead to look for the answer, only to find that the question was not quite what you expected. For example, you might be asked a question about the lengths of a dipole's elements. You quickly work out the wavelength concerned, find what you think is the correct answer and move on, forgetting that a dipole is half a wavelength and each element is only a quarter wavelength. You need to be clear which dimension the question is looking for; the total length or the length of each element. Read and read again!

More good ideas:

1. Read through *all* the questions before you answer *any* of them. This will get your memory ticking over.

2. Return to the beginning and answer all the questions that you are absolutely sure about, leaving any that are not obvious to you.

3. Your next run through will allow you to reason through the questions that you were not sure about on the first run. Hopefully, most of the questions will become clear on this round.

4. The final run is a checking round. Read through *all* the questions again, to be sure you have selected the answer you intended to. Correct any of those that you didn't. If there are still some questions that you are not sure about, guess the answers. Quite often you can eliminate two of the four answers, making the guess a 50/50 shot. No marks are lost for wrong answers.

What happens if I don't score enough?

IF YOU DO NOT pass the written examination you will need to wait a while for another attempt. You can use that time to revise the material, particularly any topics that seem to be causing you difficulties.

Whatever you do, don't give up!

Topic	Questions
The Nature of Amateur Radio and Licence Conditions	9
Technical Basics	8
Transmitters and Receivers	7
Feeders and Antennas	3
Propagation	3
EMC	5
Operating Practices and Procedures	4
Safety	4
Construction	2

Table 6: Approximate breakdown of Intermediate examination questions.

Component Symbols

Component	Unit
Resistor, fixed	Ohm
Resistor, potentiometer	Ohm
Resistor, pre-set	Ohm
Resistor, variable	Ohm
Capacitor, fixed	Farad
Capacitor, electrolytic (polarised)	Farad
Capacitor, variable	Farad
Inductor, fixed	Henry
Inductor, with core	Henry
Transformer	nil
Quartz crystal	Hertz
Semi-conductor diode	nil
Diode, variable capacitance	nil
Diode, light emitting	nil

Component	Unit
Transistor, bipolar NPN (Note Transistors can be drawn with or without the circle)	nil
Transistor, field effect (FET) (Note Transistors can be drawn with or without the circle)	nil
Earphone	nil
Microphone	Ohm
Loudspeaker	Ohm
Cell	Volt
Battery	Volt
Bulb (lamp)	Watt
Switch (SPST)	nil
Switch (DPST)	nil
Fuse	Amp
Antenna	nil
Earth	nil
Ground chassis	nil

The Intermediate Licence — *Building on the Foundation*

Syllabus Cross Reference

THIS INFORMATION is intended to help you track down the worksheet(s) that contain the training material for each syllabus item. The syllabus references were correct at the time of writing, but you should check with the Radiocommunications Agency for any amendments.

1 Amateur Radio

1a Nature of Amateur Radio
  Licence Conditions 1   p17

2 Licence Conditions

2a Operators
  Licence Conditions 1   p17
2b Messages
  Licence Conditions 1   p18
2c Location & Identification
  Licence Conditions 1   p18
2d Unattended Operation
  Licence Conditions 2   p43
2e Log
  Licence Conditions 2   p44
  Detecting Unwanted Emissions   p68
2f Apparatus
  Licence Conditions 2   p44
2g Licence
  Licence Conditions 2   p45
2h Licence Schedule
  Licence Conditions 2   p45

3 Technical Basics

3a Units of Measurement
  Multimeters   p13
  Calculating Input Power   p21
  Capacitors, Inductors
    & Tuned Circuits   p27
3b Simple Circuit Theory
  Conductors and Insulators   p5
  Calculating Input Power   p21
  Demonstrating Ohm's Law   p30
  Measuring Resistors   p25
3c Cells and Batteries
  Power Supplies   p46
3d Capacitors
  Capacitors, Inductors
    & Tuned Circuits   p27
3e Inductors
  Capacitors, Inductors
    & Tuned Circuits   p28
3f Tuned Circuits
  Capacitors, Inductors
    & Tuned Circuits   p29
  RF Oscillators   p31
  Receivers   p50
3g Transformers
  Power Supplies   p46
3h Diodes and Transistors
  Semiconductors   p35
  Using Diodes   p37
  Using a Transistor as a Switch   p42

3i Circuit Symbols
  Appendix 1   p75
3j Measurements
  Multimeters   p13
  Measuring Voltage   p15
  Measuring Current   p16
  Measuring Resistance   p25

4 Transmitters & Receivers

4a Transmitter Block Diagrams
  Transmitters   p39
4b RF Oscillators
  RF Oscillators   p31
4c Mixers
  Transmitters   p39
  Receivers   p50
4d Modulation & Sidebands
  Transmitters   p39
4e Transmitter Interference
  Transmitters   p39
4f Receiver Block Diagrams
  Receivers   p50
4g Intermediate Frequency
  Receivers   p51
4h Frequency Selection
  Receivers   p50
4i Detectors
  Receivers   p52
4j AGC
  Receivers   p52

5 Feeders & Antennas

5a Feeder Basics
  Antenna Feeders   p56
5b Characteristic Impedance
  Antenna Matching   p53
  Antenna Feeders   p56
5c Antenna Impedance
  Antenna Matching   p53
5d Standing Waves
  Antenna Matching   p54
5e Antenna Tuning Units
  Antenna Matching   p55
5f Antennas
  Antenna Gain   p58
  Polarisation   p62
5g Dummy Loads
  Measuring Resistance   p25
  Antenna Matching   p53
  Checking for harmonics
    & prurious emissions   p68

6 Propagation

6a Propagation Basics   p60

7 EMC

7a EMC Basics
  EMC   p66
7b Good Radio Housekeeping
  Good Radio Housekeeping   p70

7c Interference Sources & Remedies
  EMC   p65
7d Social Issues
  Good Radio Housekeeping   p71

8 Operating Practices & Procedures

8a Q Code
  Operating Practices   p22
8b Abbreviations
  Operating Practices   p22
8c RST Code
  Operating Practices   p22
8d Relative Advantages of Modes
  Other Types of Transmission   p48
8e Other Types of Modulation
  Other Types of Transmission   p48
8f Good Operating Practices
  Operating Practices   p23
8g Amateur Satellites
  Other Types of Transmission   p49

9 Safety

9a Soldering
  Soldering Skills   p4
9b Use of Hand Tools
  Project Briefing   p9
9c Working at Heights
  Antenna Matching   p55
9d Electricity
  Fitting 13amp Plug
    & Electrical Safety   p11
9e RF
  Antenna Gain   p59

10 Construction

10a Recognise Components
  Components Symbols   p6
  Measuring Resistance   p25
  Capacitors, Inductors
    & Tuned Circuits   p27
  RF Oscillators   p31
  Semiconductors   p35
  Using Diodes   p37
  Power Supplies   p46
10b Soldering Basics
  Soldering Skills   p3
10c Colour Code
  Measuring Resistance   p25
10d Practical Skills
  Building a Simple DC Circuit   p8
  Measuring Voltage   p15
  Measuring Current   p16
  Measuring Resistance   p25
  Using Diodes   p37
  Using a Transistor as a Switch   p42
  RF Connectors   p63
  Fitting a 13amp Plug   p1
10e Construction
  Project Briefing   p9
10f Frequency Calibration
  Calibrating a VFO   p33